Our Quantum Souls

William John Cox

ISBN: 9798290581606

The book incorporates information and images from the author's multimedia presentation of *The Logic of an Infinite, Quantum, and Universal Cosmic Consciousness Generated by a Static, Electromagnetic, and Living Universe of Light Within a Limitless, Dark, and Eternal Plasma,* which was screened at the Science of Consciousness Conference in April 2024. The cover photograph is a spectro-scopic image of the Vela Molecular Ridge which was captured by NASA's Spectro-Photometer for the History of the Universe, Epoch of Reionization, and Ices Explorer (SPHEREx).

Websites

WilliamJohnCox.com

Mindkind.info

TheVote.io

YouTube Channel

@CoxWilliamJohn

Also by William John Cox

You're Not Stupid! Get the Truth: A Brief on the Bush Presidency
Mitt Romney and the Mormon Church: Questions
Target Iran: Drawing Red Lines in the Sand
The Holocaust Case: Defeat of Denial
Transforming America: A Voter's Bill of Rights
Sam: A Political Philosophy
An Essential History of China: Why it Matters to Americans
The Way of Righteousness: A Revealing History and Reconciliation of Judaism, Christianity, and Islam

Mathematics and Physics

Time Travel to Ancient Math & Physics
The Work: Second Edition
A Geometrical Structure for an Infinite, Living, Static Universe of Electricity and Plasma, as Defined by Universal Numbers
Universal Quantum Numbers: An Introduction

The Gift of Mind Series

A Message of Mind: Hello, We Speak the Truth
The Book of Mindkind: A Philosophy for the New Millennium
Mind & Its Languages of Reason
The Choices of Mind: Extinction or Evolution

Companion Books

The Mathematics of an Infinite Universe
With Liberty and Justice for All

Contents

A Dedication to Our Quantum Souls 1

The Boy's Story: A Memoir 11

Alfvén vs. Einstein 49

Assumptions 55

Quantum Relativity 63

Quantum Consciousness..................71

Artificial Intelligence and Scientific Reality.......75

Choices..................83

The Age of Sophia Nous 89

Appendix A: About Quantumness..................95

Appendix B: The Ratio of Everything
to Nothingness 99

Appendix C: A True Story About the Family
of Jesus 101

Sources..................117

A Dedication to Our Quantum Souls

Nearing the end of my written work, this little book is dedicated to our quantum[1] souls, yours and mine, something we each have, a personal self, our conscious minds. The big question is what will happen to these souls of ours when our physical bodies die? I am now approaching 85 years of age and increasingly confront the odds of inevitable reality with each breath and step, yet ultimately, physical death is certain for all of us at the end of our lives. Then what?

If the languages of our mind and its physical creations are the products of our living brain, what happens when our heart stops beating, and our brain dies? Answering this enduring question of the ages is the primary focus of these thoughts and reflections about our souls, as we rethink the structure of the cosmos and the nature of our consciousness.

This is the twentieth of my print and eBooks in current publication as retrievable and revisable repositories of my incorporeal mind. They were written about matters I have thought about over the years to document my findings and conclusions, and to explain why I have done the things

---

1   See Appendix A for a brief note about quantumness.

I have and continue to do. These books and my journals, videos, photographs, and websites contain the physical backup of my mind, all this being the composite of an old soul who seeks to communicate with you.

After so many years as a wordsmith, my hands ache and weak eyes focused by lenses strain to read the monitors. Written here after a lifetime of earning my living at the keyboards and now equipped with improved technology, these words are composed by the increasingly forgetful mind of an old man seeking a digital place to catalogue his fleeting thoughts and fading memories. Yet inside remains the emotional spirit of a curious four-year-old farm boy who continues to seek answers to problems and to imagine solutions.

Our Quantum Souls. As near as I can tell, we *are* our individual minds, our very own person, generated by the twin hemispheres of our brain as we talk to ourselves with learned languages. Each of us is uniquely and individually responsible for the actions, conduct, and character of our living being. We are who others see and experience in life, and what we are and do. Unseen and heard, we are also what exists within, the silent articulation of our innermost thoughts and observations, one who speaks with language while another listens, privately conversing, continually.

The classical theory of consciousness and memory has been concentrated on the synaptic connections of neurons in the brain to encode thoughts and memories. This conjecture is being challenged by the study of the quantum effects that occur within the microtubules of neurons and other cells in the body.

Our bodies, monitored by a nervous system and moderated by our brains, are composed of trillions of cells, each containing microtubules, which may become entangled in quantum superposition. Encoded with information, the cells of our bodies bond together to keep us going, to see where we are heading, to remember where we have been, and what we heard, saw, smelled, and thought while we were there. This quantum entanglement is manifested as our conscious self-awareness and our named identity in life, which unites our entire physical, mental, and social being into an integrated thinker, communicator, and creator.

Our consciousness exists throughout our bodies, within our gut and heart, and from the tips of our toes and fingers, up our spinal cord through our brain stem to our frontal cortex. Simultaneously, all the time, we communicate with ourselves—until we sleep and dream (and process, sort, and store our daily intake of information).

With intelligent life comes language and awareness of self and surroundings, including the movement of the moon with her tides and months, and our mother sun providing warmth and seasons. For us as human infants, the quantum bonding of our minds occurred when we first recognized our hand and reached for the light, as we leaned away from our mother's breast. Then is when we truly become alive, and our individual quantum minds are born and begin to expand beyond our mother and family. Our quantum soul willfully makes our own way, with our mind uniquely evolving to question everything we encounter, and to physically create what we imagine in our minds.

Our living, learning, tolerant, caring, joyful quantum soul is ours to cherish and nourish forever, as our inner child wanders through the reality of our gift of life with language, adult freedom, responsibilities, opportunities, and access to resources—confronting life's challenges, asking questions, seeking answers, and creating an alternative future.

If our consciousness—the quantum whole of our soul—continues to exist once we die, where in the physical scheme of things does our wide-ranging and long-lived mind exist?

The Quantum Universe. Like the revelation of quantum consciousness, another emerging scientific reality is that the Big Bang never happened; the universe does not result from gravitational processes, and it is not expanding. Instead, it appears our physical bodies are organically and quantumly bound to our mother earth, who is quantumly entangled with her sister moon to our mother sun.

Altogether they are quantumly and electromagnetically related to our grandmother Milky Way galaxy, in a positive, infinite, living universe primarily consisting of maternal galaxies existing in an eternal plasma[2] within an incomprehensibly greater, negative nothingness.[3]

Looking out into the cosmos in every direction, in every electromagnetic spectrum with more powerful instruments, we must conclude that Hannes Alfvén and Edwin Hubble were probably right. Isaac Newton's laws of gravity should not be applied to the greater universe beyond our solar

---

2   In addition to solid, liquid, and gas, plasma is a fourth state of matter. It is a hot, electrically charged gas having a mixture of ions and free electrons in which the atoms are ionized with lost or gained electrons. In the intergalactic space the plasma consists of hydrogen atoms, free electrons, and naked protons. While the temperature of this plasma can range up to millions of degrees Kevin, its baseline temperature is 2.7° Kevin (455° Fahrenheit).
3   *See* Appendix B for The Ratio of Everything to Nothingness.

system, wherein the mutual attraction of mass objects we identify as the force of gravity is simply a local manifestation and measurement of their observable quantum entanglement.

Albert Einstein mathematically calculated Newton's three-dimensional gravitational universe using an imaginary fourth dimension of spacetime. Others misinterpreted Hubble's discoveries of light wave redshift and distance correlations as being Doppler, with the acceleration of expansion stretching the wavelength, rather than the frequency of light waves being reduced by the distance traveled through the plasma at the speed of light.

The universe beyond our solar system is not bound by Newtonian gravity, which is simply the measurable aspects of our quantum entanglement with our earth, sun, and galaxy. These are expressions of the positive, physical mass of our infinite, living, electromagnetic universe within an eternal, neutral, and conducting plasma, timelessly connected by massive flowing electrical currents, surrounded by spiraling magnetic fields and concentrated plasma, within a negative nothingness.

Invisible in the optical spectrum, are massive electrical currents that connect all galaxies together into related family clusters. Observed as filaments in radio and X-ray spectra, the currents, surrounded by the gas of concentrated plasma, can stretch millions of light years through the plasma and reach temperatures of millions of degrees Fahrenheit. This image of a simulation of the filaments that connect galaxies and clusters of galaxies was created by Volker Springel, Max Planck Institute for Astrophysics and Harvard-Smithsonian Center for Astrophysics (CfA), and Lars Hernquist, CfA.

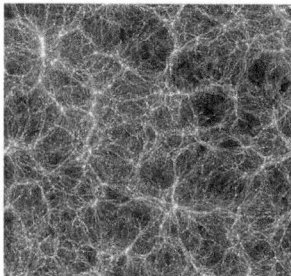

The magnetic fields that encircle these enormous and extremely hot electrical currents occasionally "z pinch," clamping down on the massive flow and causing a gigantic arc in the thickened plasma. Created is a powerful spinning plasmoid[4] at the core of a new grandmother galaxy, who generates her own arms of electromagnetic currents that occasionally z pinch, arcing and compressing her molecular gas into strings of fusion stars. Each spinning mother star generates electrical energy

---

4   If the universe is electromagnetic, rather than gravitational, then the core of galaxies is more likely to be a plasmoid instead of a black hole. A plasmoid is a coherent structure of plasma and magnetic fields. It is easily created in the laboratory and can be scaled up by simulation to galactic size. Plasmoids are at the core of attempts to build fusion generators.

in the surrounding plasma warming her planetary children, the rare ones of which attract water, sprout organic life, and grow intelligent beings.[5]

Rather than an imaginary black hole, it is far more likely that a powerful, energetic plasmoid located at the center of our Milky Way galaxy connects through massive electromagnetic currents to family clusters of galaxies infinitely—out into the eternal plasma, forever in every direction.

Bonding With the Universal Quantum Consciousness. If in fact: (1) the physical universe is infinite, and it is not expanding, as has now been shown in every spectrum by the Webb, Hubble, X-ray, and radio telescopes; (2) the quantumly related universe is manifested by electromagnetic and plasma processes, rather than by being gravitationally bound into imaginary big bang expansionism, black holes, dark matter, and dark energy; (3) there are an infinite number of family connected grandmother galaxies, each producing organic life and quantum minds that survive death; (4) then our quantum soul arises following birth and sensory awareness, and it is entangled with the scientific and historical reality of an infinite, timeless,

---

5   The lifetimes of living grandmother galaxies may be measured in trillions, rather than billions of years. Our most advanced space telescopes reveal both young and old galaxies, and our observational lifetimes may not coincide with the rare birth of a new one within our universal neighborhood.

universal consciousness simultaneously consisting of all quantum minds and their creations.

Once our infant eyes follow our hand as we reach for something, our mind is born, independent of our mother, yet we are never alone; we become an integral part of the universal consciousness, bound forever, without sin or judgment.

My mind is speaking to you in words of common language to convey complex concepts concerning what happens to our soul, however defined by religion, philosophy, or science, when we die, and our bodies are buried or burned. Will our death be violent, by illness, hunger, thirst, and pain; will we be at peace, respected, joyful, and surrounded by loved ones, or will we be alone and forgotten?

It is to you, the curious reader, that these words are directed, to your soul who talks and listens, the bonded twins of your mind, who will continue to live after the death of your body and brain. They will quantumly exist as the negative reflection of the positive physical you—as an immaterial presence, whenever your soul is thought of, and you and your creations are remembered by others.

For better or for worse, your continued self-awareness (without physical sensory perception or pain) becomes your personal heaven to enjoy or hell to be suffered, as you are remembered by others and by your contribution to historical reality. Will your work and creations endure; will your progeny prosper and care? Will your quantum soul continue to exist without physical form timelessly in the invisible light of the "Mind Field" of nothingness, as an unfathomable scientific reality, something to be anticipated, when breathing one's last?

These words are written by this solitary soul, who started as a wandering farm boy telling others about the things he saw (including lights that appeared in the sky at night as he lay on top of a hay stack counting the stars), where he was going to go, and what he was going to do when he got there.[6]

---

6   While it may be unusual to place biographical information about the author following this summary dedication and before the substance of the book, this analysis of the scientific revolutions of cosmology and consciousness requires an insight into the soul that synthesized and consolidated these startling revelations. In the following autobiographical stories, to the extent they are relevant to a comprehension of the work presented, I will relate the evolution of my mind, and some of the quests it has undertaken that led to the universal language of artificial intelligence and the geometric and mathematical structuring of the cosmic nothingness.

The Boy's Story: A Memoir

The story of my soul begins with life on a Texas Panhandle dryland cotton farm, being delivered onto my grandmother's bed in February 1941 by a doctor who had driven out from Lubbock. The birth was reported on a certificate as "Baby Boy Cox," but my father called me Billy Jack when he arrived in our old pickup truck to take us home. As the eighth living child of eleven birthed by our mother, the farmhouse we shared was farther out in the country.

The wood frame house was lit by kerosene lamps; a windmill pump provided indoor cold water in the kitchen sink, and food was cooked on a wood stove fired with mesquite limbs cut with saws and axes from wild patches by my father. He

farmed our 200 acres of land with workhorses, until he was able to buy a tractor, equipment, and fuel during the war.

The outhouse was across the farmyard, and Saturday baths with lye soap were in a washtub in the kitchen, starting with the oldest. Water was heated on the woodstove, and the spilled water was mopped around the linoleum floor with a broom and out through holes drilled around the edges. Our clothes were scrubbed on a washboard, starched, and pressed with stove heated flat irons.

The above photograph shows Mother at the kitchen door of the farmhouse, and the following is of me standing on the back of one of our workhorses.

For me, self-awareness arose on a cold day just after Christmas in January 1946. I was not yet five, and our 46-year-old mother had gone away on Sunday afternoon to the hospital for female repair surgery the next morning (accompanied by the experimental implant of radioactive material into her ovaries to "push her into menopause"). I last saw her as she was filling our pickup truck with gasoline, and she told me to be a good boy until she returned.

When our father left to bring Mother home on Friday morning, he told my sisters to clean the house, and they told me to take a nap. I was still in bed when he returned and listened through the closed door as he told my sisters that our mother was not coming home, that she had died that morning. A blood clot had broken away from the unhealing radioactive surgery site and traveled to her brain.

Hearing my sister's screams of anguish and pain, and the helpless cries of my brothers, I lay there all that day and night, pretending to be asleep, wondering what it meant that Mother was never coming home. How long was never? Thinking about who would cook for me. I became aware of my need to care for myself, and I began to think about breakfast the next morning and how I could ask a sister to make pancakes like Mother did.

Reassuring myself, talking to myself, shaping my future, I became my own lifetime best friend during that long and sad night, and the two of us have been self-aware and self-directed ever since, agreeing and working together to resolve the problems presented by life, trying our best to always be a good boy, missing Mother every day.

It was said that I did not cry at my mother's funeral; I was overpowered by the magnitude of the family's sorrow over the sudden death of she, whom everyone, including our father, respectfully referred to as "Mother." She had held her family together throughout the leanest of the Great

Depression years, aided by her widowed mother, who lost her eldest child.

But I did cry at my father's funeral just five years later when I was ten. He had collapsed following the East Texas funeral of Mother's mother, our last grandparent. At ten years old, I was unsure if my tears were real, but I knew I was even more alone and on my own.

My father had shared what he had time to give during those harsh years of drought and sandstorms, especially what he knew about the movement of the earth, moon, Venus, and sun. He also taught me how to read his dime western novels before enrolling me in the rural school when I was five, conspiring with the principal to use his August birthday to make me old enough.

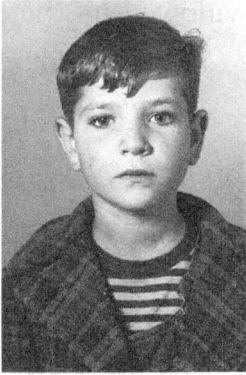

My life was profoundly lonely and boring, except for the books of the small county extension library in a room behind the café across the highway from the school. After my father's death, I read his bible and secret Masonic texts, as I thought about becoming a minister. I found solace in the large painting of Jesus praying in the Garden of Gethsemane that hung above the altar in the Methodist Church we attended and where I was first baptized. I have always felt a comforting presence of that gentle Jesus beside me, wherever I've gone and whatever I've done.

Living with my brother, I became a chronic runaway teenager from the hard life of cultivating the farm where I had been raised. These are images of me plowing the fields and raising hogs for show and market.

Hitchhiking down the highways and "borrowing" my brother-in-law's Buick Century Rivera in the summer of 1957 to tour the more pleasant Hill Country of Central Texas, my road trip was cut short when I crossed paths with a pair of Texas Rangers. I was arrested, made a ward of the court,

and allowed to attend the New Mexico Military Institute, in lieu of reform school. I graduated and spent four years in the Navy as a Hospital Corpsman before becoming a police officer in

1962 as part of the "New Breed" movement to professionalize law enforcement. To avoid fights, I trained and worked with a young male German Shepherd K9, named Lobo.

In a justice system career that spanned more than 50 years, I went on to write the Policy Manual of the Los Angeles Police Department containing its principles and philosophy of policing,[7] and then the role of the police in America for President Nixon's National Advisory Commission on Criminal Justice Standards and Goals, while attending law school in the evening. Upon graduation and passing the California State Bar exam, I spent a year working for the Justice Department in Washington, DC implementing national police standards.

Swearing to support the Constitution, I was administered my attorney's oath by retired Supreme Court Justice Tom Clark in his chambers, before being appointed as a prosecutor for the Los Angeles County District Attorney's Office in the mid 1970's.

---

7   LAPD required my birth certificate, and I learned in 1968 that I had never been officially named. Realizing that Billy Jack was a nickname, I named myself William John on the certificate.

Trying increasingly serious felony jury trials in the superior courts, I was reassigned to a rotation of prosecuting youthful offenders. Having been a product of the juvenile justice system, I found little joy in arguing for the punishment of children. I was also teaching evening criminal law classes to young people at a local community college.

At this time, I lived on the South Bay beaches and spent evenings watching and filming ocean sunsets with a Super 8mm camera and scoring movies with the music of the times. I reoriented myself to the movement of the earth, moon, and planets, as I contemplated the purpose and direction of my life of public service.

I had lost my trust of organized religions and faith in a judgmental God over the years as I read science, philosophy, and history. Along with that increased knowledge, however, came a reassured comfort in the reality of an historical Jesus, who lived, taught, and was murdered for what he was teaching—the essence of which has endured at the core of Christianity and Islam.

I found a renewed belief in, and spiritual dedication to discovering and revealing the true message of Jesus, who must have indeed prayed alone in the Garden at the Mount of Olives before going to symbolically cleanse the Jerusalem Temple of Sadducean pollution on behalf of the alternative Zadok priesthood.

As a priestly leader of the Osim[8] in their war against the Roman occupiers (and their lackeys, the priestly Sadducees, the lawyerly Pharisees, and the despised Herodian kings), Jesus knew the result would be the agony of his crucifixion. He would die alongside other Zealot fighters whom Jesus and his brothers led in the war of the People to be left alone in liberty to simply live according to the ancient pledge of Abraham to sojourn in the Land with righteousness and justice. (Genesis 14-19)

I could no longer accept most of the New Testament writings by Paul, who was both a

---

8   Commonly known as the Essene, or the Poor, the followers of the Way of Righteousness most likely referred to themselves as the Osim, which is a Hebrew acronym for "Doers of the Law." In Romans (2:13) Paul says that the doers of the law will be justified. Members were also known as Zealots, who were zealous for the law and fanatically opposed to Roman occupation.

Pharisee and a Roman collaborator, and whose mother was a Herodian. But I continued to rely on the books by Jesus's brothers, James, Jude, Hebrews, and the Gospel of John, as validated by the Dead Sea Scrolls. These works are also clarified by the Gnostic Gospels, including that of Thomas, which reflect the essential spiritual teachings of Jesus as taught by his favored companion, Mary Magdalene, who took the Way's Spirit of Wisdom westward. (*See* Appendix C.)

I considered going back to school to become a minister but concluded that I had a law degree, license, and experience that would allow me to more effectively serve the historical Jesus. Rather than becoming a preacher, I had the power of the law to file lawsuits, motions, and subpoenas in matters relating to justice and to make things happen, as Jesus would have me do.

I had kept journals over the years, and as I reviewed them, I found that I had not always been honest with myself. I discarded most of my writings, keeping only a few poems that I used as the outline of a little book on the evolution of my mind, as the two of us learned to converse more freely.

A Healthy Poem

To be what you thought,
And I wished I was,
Would be to be,
What I'm not,
Because,
I am what I am,
And not what I'm not,
But that's no reason
I can't be what I want,
For, not is now,
And then is when,
I will myself change,
Now and then,
Not to be what I'm not,
But to be what I want.

Over a period of several weeks, I engaged in lucid dreaming each night about a variety of subjects, and I awoke in the morning and immediately wrote down what was fully formed in my mind on a legal tablet by the bed.

In 1978, I published *Hello! We Speak the Truth* (*A Message of Mind*) under the pseudonym of Thomas Donn, speaking as the twin, male and female, voices within me.

The book became the philosophical guide for the life I have lived in the decades since. Freed from the judgement of a jealous creator God, and the burdens of sin, guilt, and the doubts of obsessive self-criticism, my liberated mind communicated more easily with my old best friend—the more feminine side of myself.

Acting with concern, empathy, and dedication to duty, each outside-the-box legal endeavor was followed by a lesson-learned review. Mistakes were readily acknowledged, along with a willingness to go wherever observations, education, and duty took me. The result has been my quietly and expectantly facing the challenges of each new day over decades of service, using the tools and means available to me, adapting to the needs of each new task, and doing what I could do to make a difference.

In the late Seventies, I purchased the historical landmark Skinny House in Long Beach and set up a law practice mostly for the public good, but also for the support of myself and family. Professional announcements mailed to the jurors who had sat on the jury trials I had prosecuted, and to the students I had taught, provided the client base of a successful law practice. I also received criminal defense appointments from the judges I had appeared before, and I began to primarily represent youthful offenders in serious cases, including high-profile gang and home invasion murder cases.

I became friends with the op-ed editor of the local daily newspaper and his reporter wife, and he challenged me to file a First Amendment petition in the U.S. Supreme Court in the manner of *Gideon vs. Wainwright*, complaining that the government was falling into the grips of corporations and special interest groups, and no longer represented the People who elect it.

The requested remedy was to hold a national policy referendum each time we vote for president; we the People could make our own policy, and we would elect those who will best effectuate our policy. I wrote and filed the petition in person on July 9, 1979, as a class action lawsuit on behalf of all U.S. citizens.

The Supreme Court declined to accept our case, but my journalist circle now included the

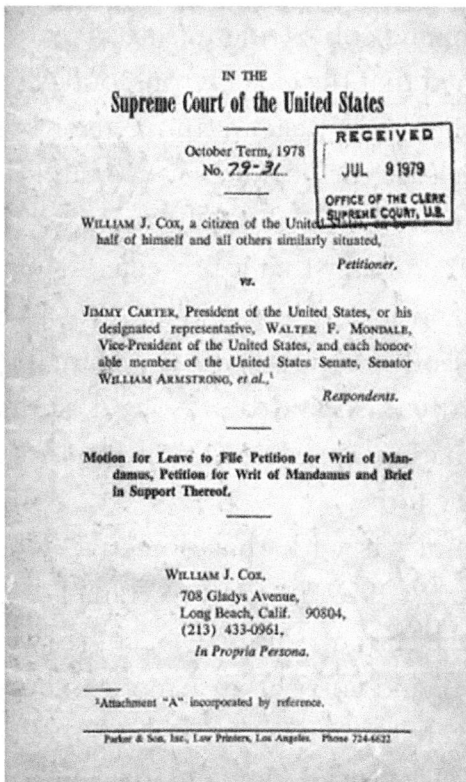

Washington DC bureau chief of the *Hearst* newspapers who reported on my petition, and the *Los Angeles Times* later said that the matter was one of the ten more interesting cases the Court did not hear that year.

We began to talk about the "hostage situation" at the U.S. embassy in Tehran, as President Jimmy Carter struggled with the difference between creating policy and personally implementing it, and candidate Ronald Reagan was challenging Carter's leadership and subverting his efforts.

I argued that the exact wrong policy was being adopted by both Reagan and Carter, who were saying "release the hostages so we can bring them home." Instead, they should firmly say, "please get out of our embassy that is protected by both international and Islamic law and allow us to continue having good relations with the Iranian people." Such conflicts should be resolved as matters of law and natural justice, without the threat or use of military force.

Thus, it was with that objective I set out in December of 1979 to travel to Iran. The plan was to go to the entrance of the U.S. embassy in Tehran and to demand entry as an American citizen entitled to exercise my right to refuge under international *and* Islamic law. I would join the hostages, at which time my journalist friends would publicize the proposed alternative policy as a peaceful resolution of the crisis.

Unable to obtain a travel visa in the United States, I flew to London to apply at the Iranian consulate there, but found it closed for the winter holidays. I was informed that it might be possible to fly to Iran from Jordan, after crossing from Israel at Amman. So, I flew to Tel Aviv and took a taxicab to a small hotel in West Jerusalem near the old walled city, arriving late in the evening, jetlagged, and falling into a deep sleep.

I dreamed I was foolishly praying to God for a "sign" as I lay sleeping on my back, and I felt a large hand take my right foot and extend it down flat. It quickly became painful, as I imagined it held by the strong hand of Saint Peter. I said to myself, "the narrow bed is short, and the sheets are tight, I have a cramp in my foot." Continuing my dream, I heard a voice say, "Wilhelm," which is how my teenage son, Steven would sometimes speak to me. But as the dream sequence went on, the voice was that of Adolf Hitler, who as a baptized Catholic, did a last minute act of contrition and begged God for forgiveness—seconds before blowing his brains out.

From my vision, it appears GOD may have both a sense of irony and humor, as Hitler's penance was to serve as a small claims judge in Jerusalem, patiently resolving the most petty and shrillest disputes of Jews forever. However, Hitler could never achieve salvation so long as there was anyone who believed that what he did was right, *and* the most horrible thing he did was the murder of the children. I awoke from my dream filled with a sense of duty to defend the children of the Holocaust.

I got up, showered, dressed and walked from the hotel before daylight into the old walled city through Jaffa Gate and up through the dark tunnel streets of the sleeping city, and out the Lion's Gate on the east. I crossed the Kidron Valley and walked up pathways to near the top of the Mount of Olives, above the Garden where Jesus had wept as he contemplated the agony of his death and the suffering of his people.

I sat down on a rock looking across the valley at the walled city. As the sun came up behind the Mount of Olives and shined down upon Jerusalem below, I heard these words within my head: "just as the same sun shines upon the roofs of the synagogues, churches, and mosques, all those within worship the same God, and there are no footnotes, asterisks, or exceptions in the Ten Commandments."

Returning to my hotel and my attempts to arrange ground transport to Amman and a flight to Tehran, I learned that once my passport had been stamped into Israel, I would be refused entry into Jordan. I flew back to London and spent a very short winter day in a hotel room watching the sun rise at the lower left corner of my south facing window, and set a few hours later at the right corner, as I wrote about my vision in Jerusalem. I flew home and continued my practice and political activities.

The Presidential Election of 1980 came down to a choice between Carter and Reagan, and finding myself opposed to the policies of both, I announced, in consultation with my journalist friends, that I was an independent write-in candidate for president. My platform consisted of a National Policy Referendum, but it also proposed a policy concept of a police response to international conflict instead of military war.

Rather than a declaration of war against a nation or people, the president would seek a "warrant" in the Congress authoring the use of force to "arrest" the named person who posed a risk of harm to our People and to those subject to his control. The identified individual would be held to answer in the International Court of Justice to defend his "government," while the victims of his misrule would not be hated or harmed. My campaign was limited to a midnight talk and music show on the local rock station, and other low-budget activities.

I was not elected, but in the days following the election (when Reagan was resting at his mountaintop ranch and the world news media was gathered in a hotel at the foot of the hill near Santa Barbara), I drove up the coast and went into the hotel cocktail lounge where the reporters could be found. We shared a few rounds of afternoon drinks, and I held a press conference during which I conceded the election and did not demand a recount.

I urged that the new president "kick a few tires" before signing off on a massive, unnecessary round of military expenditures, since the Soviets were so capable of lying and concealing an inferior force. Before leaving, I dropped off a handwritten letter personally addressed to Reagan regarding these policy issues at his press office at the hotel.

My Jerusalem vision of defending the children of the Holocaust came to reality the next year when I was contacted by Mel Mermelstein, a local businessman. As a teenage boy, he was the sole survivor of his Hungarian Jewish family who were murdered at Auschwitz in the summer of 1944, which he wrote about in *By Bread Alone: The Story of A-4686.*

A $50,000 reward had been offered by the Institute for Historical Review (IHR) for evidence that Jews were gassed at Auschwitz. The Institute, which was seeking inroads onto college campuses, wrote to Mel challenging him to present his evidence during a mock trial at their annual conference, or else his refusal would be broadcast as evidence he was lying in his book. Mel had been denied assistance by the Anti-Defamation League and the Simon Wiesenthal Center, and as a child of the Holocaust, he was looking for a lawyer to help him.

I met with Mel, reviewed his correspondence, and questioned whether anything there was legally actionable, but I agreed to think about it and to get back to him. That night, continuing my practice of lucid dreaming, I awoke remembering one of the earliest civil cases taught in first year contract law. If you mail a letter to a property owner and offer to buy the property for an exact amount, and the owner fully accepts the offer by reply mail, the contract becomes legally binding and enforceable upon postmark. I proposed to Mel that we completely accept the written reward offer from the deniers, creating a binding contract–trusting they would renege in some way, allowing us to directly file a lawsuit in the Superior Court.

Investigating, the local IHR was found to be the "scholarly" arm of the Liberty Lobby in Washington DC. With its weekly tabloid newspaper *The Spotlight*, the Noontide Press, and other publications, promoted by a massive mailing list, the organization was the most powerful radical rightwing political force in America at the time. It was all under the control of Willis Carto, the founder of the Populist Party who worshipped Hitler, and who saw himself as Hitler's spiritual successor and ruler of the United States.

Carto's publishing conglomerate peddled cultural and political hatred literature to Americans (including Oklahoma City bomber Timothy McVey), which promoted antisemitism and racial bigotry across the nation. McVey and his coconspirators used Carto's *Spotlight* telephone calling card system to avoid leaving records of their communications.

When the IHR put off Mel's claim in favor of another, my associate Daniel Mangan, whom I had earlier taught at community college, researched, and we drafted a legal pleading that not only alleged a breach of contract, but we created the tort of "Injurious Denial of an Established Fact." As an element of proof, the cause of action required that the fact be one the Court would be required to judicially notice, as a matter of law.

We then prepared a massive motion for judicial notice that included declarations from Gideon Hausner, the prosecutor of Adolf Eichman, Simon

Wiesenthal, and serious historians documenting the Nazi genocide of European Jewry and Gypsies during World War II, specifically Hungarian Jews during the summer of 1944.

Through a PERT charted schedule of an aggressive strategy of discovery, including interrogatories, depositions, demands for admissions, and motions to compel upon defaults, the accelerated matter was set for mutual motions of summary judgment in the Superior Court of Los Angeles County. The case had been followed by the *New York Times* and other national media, and on the day of the hearing in October 1981, all three major television networks were allowed in the courtroom through a pool camera.

Accepting the defendant's position that Auschwitz was simply a work camp, I argued, "But

where did all the babies go, Your Honor? This is the question. Where did the children go? They were not subject to labor. They were not available: they were not there. And they were put to death. And that is really what this case is all about."

Ruling on my motion for judicial notice, Judge Thomas Johnson held that it was simply a fact and not subject to dispute that "Jews were gassed to death at Auschwitz Concentration Camp in Poland during the summer of 1944."

In the weeks following the decision, I traveled to Jerusalem regarding some unresolved aspects of my investigation of Carto's empire. I was invited to have morning tea with Prime Minister Begin, and I shared with him my client's book and the results of our case. Begin fiercely declared, "Never Again! Never again will Jews be led like lambs to the slaughter."

Shouldering the financial debts of the case, loss of income from my neglected practice, and negative credit rating, I closed my law office and took a year off to think about other things. In a series of thought experiments, I imagined contracting the eight corner and six face vertices of a Cartesian Cube to simultaneously pierce zero, continuing into the negative nothingness. The result of connecting the 14 equal negative vertices was the *Universal Sphere*, with six great curves describing 24 equal right-angle spherical triangles, having circumferences of *Pi* times radius, with side ratios of 3:3:4. But, that and the *Universal Numbers* created to describe the geometry and to serve as a mathematical framework for the observable, infinite universe, is another story, told elsewhere.[9]

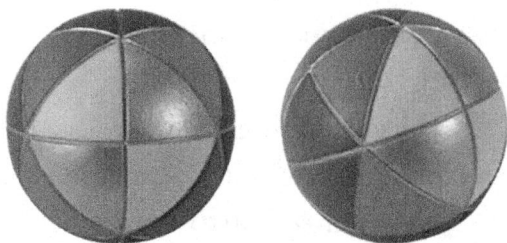

Forced to earn a living, I was recruited in 1985 by my former LAPD commanding officer and his partner who created the SWAT team concept. They established a private corporation to offer worldwide high-level security planning, armed protection, and investigation services. Providing protection for a corporate sponsor and its guests was very lucrative during the 1984 Los Angeles Olympics, as were services to master corporate contractors at the nuclear weapons production and storage sites to train, test, and evaluate the

9   Simultaneously published as a companion book is *The Mathematics of an Infinite Universe,* (2025).

protection forces that defend the weapon-grade special nuclear material. Through another corporation, they performed Energy Department background investigations of the workers at nuclear power plants. The two principals hired me as their general counsel and operations officer to coordinate and document procedures and tasks, as they traveled extensively to direct matters on-site.

Using the latest computer technology, we were able to timely produce contract proposals, staffed with vetted personnel, to quickly deploy teams as needed, conduct force-on-force exercises with lessons-learned critiques, and to produce book size, bound paper reports. With offices adjacent to a commercial airport runway, the security directors of our corporate clients could land and bring their problems to us for quick and confidential solutions.

My last act as general counsel for my principals was to negotiate the sale of their corporations to a group of investors who were assembling other related companies under an umbrella of corporate security services. I resigned but continued to consult with my principals as needed during their transition period of contractual obligations to implement our documented professional systems as the model to standardize other acquisitions.

In 1991, the Holocaust Case was presented by Turner Network Television in *Never Forget*. Payment for the movie rights enabled me to pay off the remaining debts from my prosecution of the matter ten years before.

I was able to semi-retire in a large studio apartment on Ocean Boulevard in Long Beach overlooking the Harbor. Opening an "investigative" law office and operating the latest personal computer connectivity and subscribing to the same commercial data sources we had accessed for our corporate clients, I began to offer confidential investigative and evidentiary consulting services to the large LA area law firms at a time before the Internet search engines were available.

Continuing to labor for my spiritual client, the historical Jesus, and acting on behalf of a "secret client," I signed a contract with the Biblical Archeological Society (BAS) in 1991 to publish a collection of almost 1,800 photographs of 2,000-year-old fragmentary documents found in the Dead Sea caves. Access to these scrolls had been denied to biblical scholars for more than 40 years.

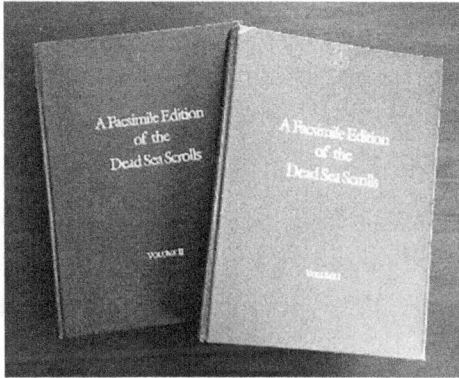

Inasmuch as the documents revealed the common roots of Judaism, Christianity, and Islam, I finally understood the meaning of the words I heard in my mind on the Mount of Olives that observed they all worship the same God.

Against my wishes, and that of the editors, the BAS publisher included a short translation of one of the scroll fragments by an Israeli professor, without consent or attribution. Acting with the support of the Israel Antiquities Authority (that claimed putative legal ownership of the scrolls after the 1967 capture of the Rockefeller Museum in East Jerusalem), the offended professor filed suit against the BAS and editors in the Jerusalem court.

I contacted Jerusalem attorney Amos Hausner, the son of Gideon, Eichman's prosecutor, who agreed to represent professors Robert Eisenman and James Robinson. The editors had written an introduction and prepared an index for the book, and both had objected, as had I, to the inclusion of the challenged translation by Hershel Shanks, the publisher.

I testified at a trial held in Jerusalem in January and February 1993, during which I related the opposition of the editors to publication of the contested translation. As the last witness and responding to the final question, I refused to identify the source of the photographs when asked and was threatened with jail for contempt of court. I have never disclosed the identity of my client.

During the trial and a rare snowstorm, I stayed at Christ Church inside the old city walls near Jaffa Gate.

Studying James Robinson's publication of the English language edition of the Gnostic Gospels and Robert Eisenman's books about James the Just and the Dead Sea Scrolls,

I discovered other related historical works that were validated by the new discoveries. Relying on the latest documentary and archeological findings, I began to research and write about the revealed history of Judaism, Christianity, and Islam. Completed in 1994, was a thousand-page manuscript entitled *Mary, Mother of Israel's Messiahs, Joshua (Jesus) The Suffering Messiah, Judah (Thomas), the Twin Messiah, and Jacob (James), the Priestly Messiah.*

Semi-retired, I once again had the time to think about *Universal Mathematics* and spent weeks at my desk doing hand calculations of the scientific values of *pi, phi,* and *e* in (16-base) *Universal Numbers*. This allowed the development of more geometric models and the identification of negative fractional geometric numbers.

After fighting off two thugs who tried to mug me with a bottle of wine one evening while walking home from the market, leaving me with even more scars on my face, I adopted and rigorously trained a young female mixed breed golden German Shepherd/Australian Dingo named KD. She became my constant companion in a working harness during those years of semi-retired living, reflection, and travel in the late 1990s.

Living in an apartment, most days I would take KD to the beach or to our local dog park where she would endlessly perform her signature trick of leaping into the air and snatching a tossed-up tennis ball, while twisting 360 degrees and landing at your feet, dropping the ball, sitting, ready to go again.

One day, a little seven-year-old girl with school braids asked if she could toss the ball. From then on, Naomi joined me whenever she was in the park with her little rat terrier. My new friend would keep KD occupied while her Buster chased the dog pack.

I assumed the girl's standoffish mother was married, until one Saturday morning at the park when she was alone, and I inquired, "Where's Naomi." Helen answered, "It's her weekend with her father," and I asked her if she would like to have lunch sometime. We did, but then some awkward months passed in Dog Park as we got to know each other, before an opportunity arose to help her with some carpentry work on a nonprofit's photography booth at the Cajun Festival. We did the job, and the next time Helen saw me at the park, she said if I wasn't doing anything the following Saturday, she would like to get together.

It was a lovely day, and I took Helen on a driving tour of my favorite places: lunch at the Los Angeles Municipal Art Gallery and Hollyhock House in Barnsdall Park, out to the coast and down to Magic Alley in Manhattan Beach, on down to the Neighborhood Church in Palos Verdes, and then around the peninsula and up through San Pedro, across the harbor bridges, and back into Long Beach. We arrived in time for a spicy dinner in Cambodia Town. We have been together almost every day since.

For the first couple of years, I operated my investigative law practice and worked on a book about *Universal Mathematics* at my desk in a corner of Helen's large condo bedroom, but before marriage, she told me I would have to get a real job. So, between 1999 and 2007, I served as a supervising trial counsel for the State Bar of California under the auspices of the California Supreme Court. I created and fielded a "Fast Track" team of lawyers, paralegals, and investigators that targeted the prosecution of attorneys accused of the most serious crimes and misconduct. Combining criminal and civil law with administrative procedures, we formulated a strategy to use the Superior Courts to quickly assume emergency jurisdiction over law practices that posed a substantial risk of harm to the public.

We were so successful that the California legislature extended the authority of the State Bar over the unlicensed practices of law operated by criminal gangs. Working with law enforcement officials, our team served court orders, seized files and bank accounts, instantly shut down the most dangerous law practices, and helped victim clients recover their losses.

Gainfully employed, with health benefits and a retirement plan, Helen and I were legally married on January 2, 2000, by our judge friend Judi Jones in her home, and the next month, we took a week off from work and traveled with Naomi to Jerusalem.

We were spiritually married on Valentine's Day, along with Naomi in a three-ring ceremony in Christ Church in Old Jerusalem, where I had stayed during the Dead Sea Scrolls trial.

Helen and I returned to work and our life together. We bought our small garden home that year and saved for our retirements, me in 2007 and she in 2009. She returned to university for a master's degree in classical fine arts and now works in a studio near the Long Beach Harbor, while I continue to write in my study and to contemplate life in her gardens.

Having published my first political book, a brief on the George W. Bush presidency and the Global War on Terrorism before my retirement, I primarily began to write about politics and philosophy. Promoting a peaceful political evolution, a Voter's Evolt, I was able to publish articles on a variety of Internet political sites. I also wrote books on *Transforming America: A Voter's Bill of Rights*, *An Essential History of China: Why it Matters to Americans*, and *The Book of Mindkind: A Philosophy for the New Millennium*. However, I found it difficult or impossible to obtain an agent or publisher for my books under normal processes.

Around Thanksgiving in 2014, I was hospitalized for a week with full-body sepsis, during which multiple antibiotics were administered intravenously, as my body struggled to survive. Alone during the first night, I was faced with the traditional near-death experience, with a lighted doorway appearing above me, inviting me to enter. I was tempted to see what lay beyond but chose to continue with life. Leaving the hospital in a walker, and realizing my years were numbered, I began in 2015 to self-publish the various books I had completed, and the ones I have written in the ten years since.

In 2017, I rewrote the earlier manuscript about *Mary, Mother of Israel's Messiahs* and incorporated the spiritual aspects of Kabbalism, Gnosticism, and Sufism into *The Way of Righteousness: A Revealing History and Reconciliation of Judaism, Christianity, and Islam*. The book was completed, except for an epilogue, but remained unpublished until after my son, Steven and I traveled to Israel in June 2019, for him to photograph the relevant locations (some of which are included in this book). I published *The Way* during COVID in 2020, along with *The Choices of Mind: Extinction or Evolution?*.

In the early days of the pandemic my friend Brian Gonzales programmed a computer application for me that allowed conversion of base-10 numbers to *Universal Numbers* and to make

calculations in that language. I was able to construct matrices of *Universal* base 100 (256) and the corresponding negative fractional geometric numbers. With these additions, I published *The Work: A Geometrical Model of the Universe, as Defined by Quantum Numbers, with the Quantification of Pi, Phi, e, and i.*

I have continued to petition the U.S. Supreme Court on behalf of all Americans, asserting the

---

Supr

RECEIVED
SUPREME C
POLI
2018 FEB 28 A 10:41

WILLIAM JOHN C
a Citizen and Person of the United States of America, and
a Citizen and Resident of the State of California

Petitioner,

vs.

The Government of the United States of America

Respondent.

A Declaration of the Personal Rights of Liberty
and Withdrawal of Consent to be Governed;
Motion for Leave to File Petition for Writ of Mandamus;
Petition for Writ of Mandamus Against a Corrupt,
Ineffective, Unrepresentative, and Threatening
Government, and Brief in Support of the Remedy of a
National Referendum Regarding
The Voters' Bill of Rights; Request for Judicial Notice; and
Request for *amicus curiae* Briefings.
Supplemental Filing Regarding:
The People's Inherent Personal Rights of Liberty
Are a Restraint Upon the Respondent; and
The Residual Rights and Powers of the
States United of America Are Transcendent.
Provision of Extraordinary Notice.

In the Name of The American People

WILLIAM JOHN COX

People's retained Rights of Liberty under the Ninth and Tenth Amendments of the Constitution to compel our federal government to be more responsive to our needs. All to no avail, nor was my *amicus curiae* brief asserting those original rights of liberty on behalf of American women, pleading for a rehearing on their loss of inherent reproductive rights.

Having exposed my quantum soul, and to escape the current, distressing, worldwide political turmoil, I will return to the geometry and physics of the cosmos and the mathematical languages of my mind. We will next consider the reality of a magnificent, infinite universe of light where our living souls can timelessly soar forever, pausing whenever we are remembered and wherever our progeny thrives in joyful learning, creation, and exploration.

Alfvén vs. Einstein

As a young trial lawyer working in the public interest, I would occasionally make up a cause of action or file a spurious motion to achieve a just result, when there was no established remedy at law. Thus, proffered here is a fictional legal case between conflicting theories of universal cosmology proposed by two physicists last century, one based on creative mathematics, and the other on observations and experiments.

The German mathematical physicist, Albert Einstein (1879-1955) modified Newton's laws and mathematically imagined the circumstantial evidence of a physical universe (telescopically limited at that time to the Milky Way stars) that resulted from the attractive force of gravity. Allegedly, gravity is a distortion of an imaginary mathematical fourth dimensional spacetime by the energy and momentum of mass.

When the mass of the three physical Cartesian dimensions is multiplied by spacetime, calculated by the speed of light squared, new, better, and more expensive ways to kill People appeared, along with the ability to escape the gravity of Earth and to see beyond the horizon. Einstein was awarded the

Nobel Prize in Physics in 1921 for his services to theoretical physics.

The Swedish plasma physicist *and* electrical engineer, Hannes Alfvén (1908-1995) studied the direct evidence of the magnificent, colorful auroras that light up Scandinavian nights, *and* he worked in his laboratory to replicate and harness the power of solar electromagnetism. For his theory of magneto-hydrodynamics in an electromagnetic plasma universe, Alfvén was the Nobel Laureate in physics in 1970; however, his work was and continues to be disparaged by gravitationally bound and mathematically creative physicists. Alfven's work has been ignored and its academic study discouraged—until now when his vision has been brilliantly vindicated by our improved observations of the cosmos, for all to see.

Courageously, others—equally well trained, but with less support and peer respect—have continued to view the heavens with what little telescope observing time is allotted them, seeing the cosmos for what it is: a living, electromagnetic, nonexpanding, infinite universe of related grandmother galaxies. Instead, the false physics of gravitational relativity has been criminally misused for weapons of war and genocide for more than a century of mindless, wanton, wasteful, industrial militarism.

As an independent expert witness in the dispute between the two theories, astronomy professor Edwin Hubble (1889-1953) used an improved optical lens telescope to reveal the faint, wispy light of *nebulae* to be individual galaxies like the Milky Way. This led him to the direct evidence of a redshift-distance correlation of galactic objects, but *he did not perceive an expanding universe.* His own redshift findings, however, were misconstrued as

circumstantial evidence of Doppler redshift reces-
sion by the gravitationalists to prove their mathematical, religiously inspired Big Bang, which
Hubble did not accept.

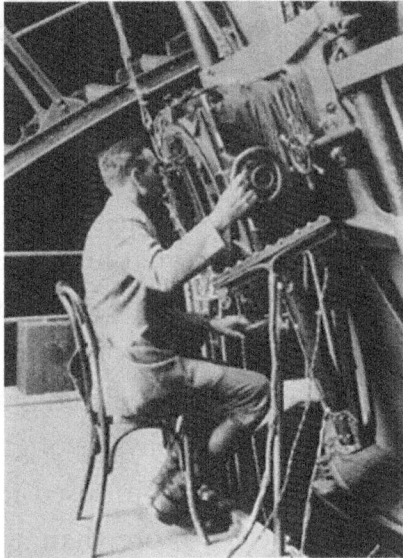

In addition to the compelling scientific direct evidence that has been subsequently discovered, also pled is the ancient law of equity.
Inasmuch as the honest life work of Professor
Alfvén has been falsely and wrongly accused of
being pseudo-science for fifty years, any circumstantial evidence supporting both irreconcilable
cosmological theories would dictate the standard
instruction by a presiding judge to a sitting jury.

If the circumstantial evidence can indicate either possibility, jurors are compelled by law and justice to render a posthumous verdict of innocence for Hannes Olof Gösta Alfvén, validating his vision of our living, electromagnetic universe of light and mind, and establishing the truth of scientific reality.

Assumptions

Combining the latest revolutionary science from both cosmology and consciousness, *Our Quantum Souls* can be summarized by a series of related assumptions relevant to the scientific and historical reality of planetary life and minds on Earth.

Based on the overwhelming observational telescopic evidence, now confirmed at every spectrum and at every distance, we can reasonably assume that a "Big Bang" did not occur 13.8 billion years ago; *that* the universe is not expanding; and *that* it is static and infinite. If true, further rational assumptions can be made.

Assume that the observable physical universe of mass and light is not based on or derived from the force of gravity, which is an archaic, mathematically measurable, manifestation of the quantum relationship between mother stars and their planetary offspring.

Assume that the infinite, quantumly related universe exists electromagnetically within an eternal, neutral, and conducting plasma that produces rarely born, but long-lived, grandmother galaxies who give birth to daughter stars, who in turn become mothers of the rarest of all physical phenomena:

planets of earth and water, with a large moon, and organic life.

Assume that the surrounding positive physical universe is infinite without beginning or end, with grandmother galaxies appearing in the distance as far as we will ever be able to see in every direction with our telescopes; *that* all galaxies are quantumly related and bound together by electromagnetic filaments, coexisting within an eternal plasma; and *that* every particle of the plasma and the living, infinite, positive universe is encompassed by an incomprehensibly greater negative nothingness.

Assume that an electromagnetic wave of light lengthens by distance traveled at the speed of light through the plasma of the intergalactic medium, as its frequency is reduced by its serial collision, absorption, and retransmission by vibrating free electrons in electrostatic plasma crystals; *that* the collective loss in frequency energy from retransmissions is observed as the cosmic microwave background (CMB) radiation, and *that* the redshift of light is correlated by the Hubble Constant with the distance a wave of light flowed through the plasma (the number of its electron retransmissions), from origin to observation.

Assume that the positive, infinite, physical, living universe exists simultaneously everywhere at the same moment, as time cannot exist in an infinity without a beginning or end, nor when traveling at the speed of light; and *that* minds create time as a tool when observing and calculating matters such as acceleration, movement, and spin, and when recording the scientific and historical reality relative to existing planetary life and mind on Earth, with its sequential ages of fire and ice.

Assume that the observable universe largely consists of endless, long-lived grandmother galaxies powered by massive electromagnetic plasmoids at their cores, some emitting spectacular gamma-ray jets across thousands of light years, and all visibly creating daughter stars along their flowing electrical arms, instead of lurking as insatiable gravitational black holes that continually consume stars until they gobble them all up—in a nightmarish and mathematical fantasy.

Assume that galaxies are connected to their relatives by optically invisible, massive, extremely hot currents of electricity encased by spiraling magnetic fields and molecular gas, observable in radio and X-ray spectra, that ceaselessly flow through the neutral and conducting plasma; and *that* these currents attract concentrations of H_2 molecular hydrogen gas and H_1 dust around them.

Assume that galaxies and quasars are created when a constricting, magnetic "z-pinch" occurs along one of the massive connecting electrical currents, and a stupendous arc occurs that creates a gigantic spinning plasmoid in the concentrated molecular gas, with the resulting formation of stellar mass, earth and water planets, large moons, and organic life; and *that* ultimately appearing is the rarest flowering of living grandmother galaxies and their daughter stars—intelligent life, which evolves into quantum minds with language, who become aware of themselves and their surroundings.

Assume that grandmother galaxies are surrounded by concentrated clouds of electrically attracted hydrogen dust and molecular gas, shepherded by galactic magnetic fields; *that* fusion stars are born when z-pinch arcs occur along the flow of her electrical arms, instantly compressing the surrounding molecular gas; and *that* the resulting daughter fusion stars become mothers who provide nursing energy to their quantumly bound planets by spinning in the electrically attracted and concentrated plasma, as powerful magnetic generators of electrical energy, which we see on Earth as the lovely auroras above the poles.

Assume that conscious mind is the product of quantum processes that naturally take place in cells throughout the human body and brain; *that* mind expands beyond the body, as it grows with a questioning awareness of self and surroundings; and *that* the emerging, conversing, quantumly bound twin mind entangles and shares its thinking and creations with a wise and caring universal consciousness, which is as extensive as an infinity of accumulated knowledge and wisdom, along with the treasures of every unique creation, each an integral part of the whole.

Assume that universal consciousness is experienced as a nonphysical (and non-metaphysical) Spirit of Wisdom; *that* she embraces and comforts our quantum soul as a motherly presence, loving and caring unconditionally, without judgment or interference.

Assume that any extraplanetary contact would collapse the waveform of humanity into a single result, a probability of one, instead of an infinite number of creative alternatives.

Assume that, relative to the magnitude of the positive eternal plasma and infinite universe that produce an infinity of earthly gardens with an infinity of quantum minds, the enveloping negative nothingness is incomprehensibly greater;

that an equivalent, infinite motherly mind must also exist as a scientific and historical reality; *that* she is a collective consciousness consisting of *everything* the liberation of minds from instinctive intolerance has ever created; and *that* her knowledge of scientific and historical reality exists as a continuous and simultaneous narrative within a timeless infinity of unique factual stories.

Assume that the negative nothingness that encompasses every minute particle of the positive physical universe and plasma is the realm of consciousness; *that* this Mind Field holds our memories and our initiative, purpose, and creative imaginations; and *that* it contains an equivalent negative reflection of our positive quantum soul, who we are, what we do, and why we do it.

Assume that the I and me—we who speak to ourselves within all of us, constantly, are quantumly bound together in life; *that* we are manifested by our character and name in society, and by our role and influence in the lives of others; *that* we become our integrated quantum soul—who we are and how we will be remembered; and *that* our personal, individual soul, entangled as it is with the whole, has the innate capacity to encompass everything we can comprehend or imagine, including the timelessness of an apparent infinity, without beginning or end.

Assume that our quantum soul survives the physical death of our body with the loss of all sensory input, perceptions, or pain, leaving us alone and adrift with the images and memories of our lifetime, narrated by the languages of our mind; and *that* the negative of our positive physical lifeform continues to exist in the infinite, invisible light of the Mind Field, joined whenever we are re-membered, or thought of, by the minds of others, one way or another, or never at all.

Quantum Relativity

As we look out into the night sky in every direction, we see points of galactic light to which we are connected by continuous currents of neutral light waves flowing through the neutral plasma of intergalactic space, timelessly, from source to receiver.[10]

When the light waves from two distant galaxies line up in the above Edwin Hubble Space Telescope Wide Field Camera image, the closer galaxy is seen in the bright center foreground surrounded by the orb of its gaseous circumgalactic medium. Some of the light waves from the more

---

10   Photons, the quanta of light, appear to have infinite lifetimes. Once emitted as a wave of light, they will effortlessly flow as a continuous current of electromagnetic radiation through the negative nothingness until they impact something in the plasma.

distant galaxy are absorbed or scattered by the core of thickened plasma, with the remaining currents flow around the thinning edges of the spherical lens of concentrated molecular gas and hydrogen dust. The electromagnetic currents from both galaxies flow simultaneously at the speed of light, timelessly, from source to receiver.

Viewing deep infrared images of the universe by the James Webb Space Telescope, light from stars and galaxies can be seen coming from every direction. But what we do not see and cannot measure are all the other myriad currents of light waves that are not coming directly toward us in our line of sight, which pass by and escape our observation.

In a thought experiment, let us imagine an elemental (3^3) 27 square meters of nothingness existing in the middle of the intergalactic nowhere. Appearing black and empty, the structure contains an electrostatic crystal consisting of 27 evenly spaced, neutral, vibrating hydrogen atoms surrounding (3^2) nine negatively charged, vibrating free electrons, and (3^2) nine positive vibrating, naked protons. With the magnetic attraction of particles with opposite electrical charge and the repulsion of those with the same charge, the atoms and particles are all naturally organized and vibrating within the structure in electrostatic harmony.

It is within these basic crystals of plasma in the intergalactic space that light waves occasionally strike vibrating free electrons. Impacted electrons absorb the photons of light waves, recoil and then re-emit the photons in the forward direction along the path of recoil, with a tiny loss of energy, as the electrons rebound back into electrostatic balance.[11] This loss of energy is observed in the universe as CMB radiation.

---

11 Because the unbound free electrons exist in an electrostatic relationship, they rebound to their previous location, and the emitted photons are not scattered by the Compton effect as they would be otherwise. Photons that strike hydrogen atoms are absorbed and excite the electrons to a higher orbital. Photons that impact naked protons transfer some energy to the protons and scatter with a reduced frequency.

Other than the observable currents of light waves emitted by stars and galaxies and captured by the lens of our telescope, we cannot see the crisscrossing of light wave currents flowing in all other directions from every imaginable distant source through the nothingness.

Were we able to observationally capture a freeze-frame image of the fleeting presence of all transient currents of light flowing through our imaginary intergalactic structure, collapsing all wave forms simultaneously, the 3^3 square meters of black nothingness would become incandescent with light and energy.

These intergalactic currents of light waves have no electrical charge and do not interact with each other, yet they fill our imaginary structure of nothingness with their abundance. We cannot observe the waves because they are not flowing directly at us; however, their currents of light through the plasma simultaneously exist as a matter of scientific reality, invisibly filling the nothingness with their coexistence and energy.

Thus, continuing the thought experiment from outside the box with invisibility uncloaked, the negative nothingness can now be seen as brilliantly illuminated by innumerable, invisible electromagnetic currents simultaneously flowing from every direction through the dark. They constitute a field of invisible, neutral light energy that surrounds the electrostatic plasma crystals of neutral hydrogen atoms, free electrons, and naked protons.[12]

Instead of darkness, when viewed as a whole, the nothingness around the physical particles of the universe is ablaze with dazzling light.[13] It is the combination of this invisible transient light within the nothingness that may constitute the neutral field through which the waves of light flow and wherein the incorporeal universal consciousness exists. Comprehension illuminates and surveys the limitless nothingness, which encompasses everything physically identifiable, infinitely.

Inasmuch as the Webb Space Telescope can image candidate galaxies like the Milky Way near the Big Bang distance of approximately 13.8 billion light years, the light wave currents of our grandmother galaxy must also extend that distance in every direction, simultaneously. Thus, in scientific reality, our Milky Way galaxy quantumly occupies a geometric, spherical space in the negative nothingness that is at least that large, along

---

12   Imagine also that the three-cube structure simultaneously contains the existence and flow of trillions of virtually invisible, weightless, and chargeless neutrinos from every direction, which may be an element of the field.

13   This might explain the common image of a lighted doorway in near-death experiences.

with all other galaxies, as they all overlap and relate.[14] Therefore, *the light currents of all galaxies must coexist as an infinite, neutral, invisible field occupying the negative nothingness.*

Although its existence can only be imagined, the negative nothingness of the Mind Field may also be the location of our creative quantum minds, eternally entangled with the infinite universal consciousness following birth and self-awareness.

The Big Bang theory was an entirely creative, counterintuitive, gravitational universe created by mathematical theorists that has never been supported by observations, and which required the repeated, imaginary modifications of inflation, dark matter, and dark energy to explain the contradictory results of our improved vision. Even Edwin Hubble, the discoverer of redshift-distance correlation, disagreed that the redshift was Doppler, proving an expanding universe. Instead, redshift is a simple, accurate measurement of intergalactic distances traveled by light waves through the eternal plasma of an infinite, static, and nonexpanding universe.

The science fiction of mathematically reversing the imaginary expanding universe back to the

---

14   Ultimately, all light waves impact something in the plasma that absorbs and scatters them, which, depending on its luminosity, imposes a limit on the radius of a galaxy's orb of light.

singularity of a Big Bang (which was first calculated by a Catholic Jesuit priest) was something the creationist religions embraced, as though almighty GOD snapped his fingers and miraculously created the universe in an instant from nothing. This is not, however, what we observe as we continue to scientifically improve our vision across all spectra into the distant past.

The broad range of newly discovered evidence resolves the scientific dispute of *Alfvén vs. Einstein* beyond any reasonable doubt, and the verdict is in. The physical universe is infinite and static; it is not gravitational and expanding.[15] The physical living universe is quantumly related as manifested by its natural electromagnetic processes within the electrically neutral and conductive plasma, and it eternally exists without beginning or end, throughout the Mind Field of cosmic nothingness.

---

15   Mathematical reconciliation of the laws of gravity and electromagnetism as a manifestation of quantum relativity can be found in the remarkable similarity of their basic formulas, as both follow an inverse square law by which the force decreases proportionally to the square of the distance between two observed objects. Both can also be described by Gauss's law that relates the flux of a force through a closed surface to the total mass or charge enclosed.

Quantum Consciousness

There are numerous theories as to the origin of consciousness, and classically most are associated with the brain and are focused on the synaptic connections of its neurons to encode current thinking and long-term memories.

Consciousness may originate in an area of the ancient brainstem associated with arousal and wakefulness, which if interfered with can disorient consciousness, and it may manifest itself as higher order thought in the prefrontal cortex. Or consciousness may simply emerge from the integration of all information within the brain, relating to the amount of information, rather than the way it is processed. Instead of any one theory, scientific reality may include all of them, as they functionally coexist and collaborate to produce the mystery of consciousness.

Most troubling is the identification of the precise physical location and manner that long term memories are encoded. Where in the physical body of a composer is the full score of an hour-long symphony, with the separate musical parts for all instruments in the orchestra recorded, and how? What is the source of the popular songs and voices of others we all hear in our minds?

The "Hard Problem" of the scientific study of consciousness is the identification of the objective physical processes associated with the subjective nature of the observations and experiences of our minds.

Challenging the classical view is the model proposed by Sir Roger Penrose and Dr. Stuart Hameroff that consciousness arises from quantum superposition and computation in the microtubules of cells, particularly neurons. The microtubules have been shown to exhibit large-scale quantum resonance and quantum vibrations, which correlate with brain waves and consciousness. They can absorb and transmit light, and they can be controlled with light. (Wikipedia: Image of a microtubule.)

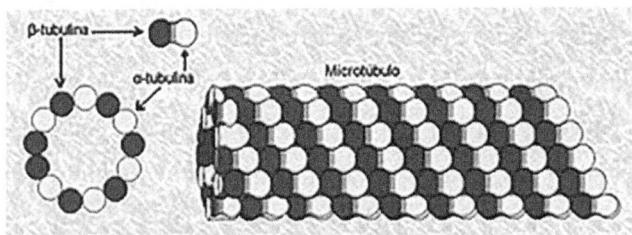

Electromagnetic processes within the microtubules of brain neurons and cells existing throughout the body, appear to produce quantum effects in the timeless transmission and storage of information that entangle the microtubules in a superposition. The entirety has evolved to encode the knowledge, memories, and language required to generate the waveform of a non-physical mind that extends beyond the brain, creating a self-aware consciousness that uses language to debate its observations and conclusions with itself.

With awareness arising after birth, an individual's mind migrates beyond the physical brain and organic body that generates it, and it limitlessly expands to the extent it can reliably observe and comprehend reality, and to the degree it changes its circumstances with its historical creations. A mature mind becomes aware of itself and surroundings. It reaches out to others to peacefully modify their shared environment and society, and to create an alternative future.

Once generated, your mind entangles with the motherly, universal consciousness and her infinite store of knowledge, wisdom, and the scientific history of all creations of mind—she who exists, without judgment or interference, everywhere.

The non-metaphysical, scientific reality of cosmic quantum consciousness (as expansive as the collective minds, knowledge, creations, and wisdom) is coexistent and coequal with the negative nothingness that envelopes the positive, infinite physical universe and plasma.

To the extent you can comprehend the existence of a timeless, infinite universe of living grandmother galaxies generated by electromagnetic processes within an eternal plasma, enveloped by the Mind Field of a far greater nothingness, the scientific reality of your quantum mind becomes infinite in its reach.

Artificial Intelligence and Scientific Reality

Because there is so much more to know than we are presently capable of comprehending, our ignorance creates tension between our deeply felt philosophical and religious beliefs and the reality of science—as we reflect upon the violent twentieth century of industrialized wars, and the criminal abuse of science complicit in the slaughter of one hundred million people.

We gaze into an abyss witnessing the horrifying spectacle of last century's deadly history repeating and magnifying itself: from the genocidal destruction of Warsaw to Gaza, from Germany's invasion of Ukraine to Russia's, and from incinerating the populations of Japanese cities to test atomic devices, to using computers and artificial intelligence (AI) to target missile and drone attacks in regional wars that threaten to go nuclear—all of which risks human annihilation and the loss of everything that is dear.

Given our critical need to know what we don't know, and must learn quickly, we have to balance the risk of trusting nonhuman artificial intelligence with the scientific reality that correctly instructed, AI can record, recollect, and correlate data, seeking contradictions, far beyond human capabilities to comprehend. Once safely mastered, artificial intelligence can produce new, presently unimaginable, useful devices for the common good of the People.

Instead of being slaughtered by old dumb weapons of war, our children may use artificially intelligent machines to spin free of the quantum bonds of gravity[16] and to soar throughout the Mind Field, along streams of electromagnetic energy connecting and powering grandmother galaxies and their starry filaments, timelessly, at the speed of light.

Artificial intelligence and machine learning are tools, like the ones we have on our garage workbench, but only fools blindly trust their power tools without first protecting their fingers. AI can be misused to improve nuclear weapons, ICBMs, drones, and the computers and satellites that guide

---

16 If we view the *gravity* we experience as we jump up and down as a measurable manifestation of quantum relativity, instead of its being an undetectable *force* of mass that attracts other mass, we may profit from discoveries that quantum entanglement can suffer sudden death at specific temperatures of heat or upon extreme acceleration. These insights may encourage the creation of transportation machines to break quantum bonds and to spin free of planetary and solar gravity. As a thought experiment, imagine canceling angular momentum and time by the high-speed counter-rotation of disc clocks.

them, all of which can be used for war or peace, depending on the purpose of their programming.

The successful application of artificial intelligence depends on the authority and intent of those who master and instruct AI. The nature and governing policies of a society provide the motivation, purpose, and bias for critical instructions and the knowledge base and cultural clues required to effectively and safely train machine learning.

The mastering and instruction of AI must be governed by the reality of science, rather than philosophical beliefs or religious faith.

Augmented by AI and borne on the wings of tolerance, the untroubled minds of our children will be empowered by the reality of science to explore our infinite, nonexpanding universe.

In this alternative, achievable reality, our children will be endowed with the gift, comfort, wisdom, and reach of our quantumly bound minds, and they will be liberated to joyfully explore our observable cosmic neighborhood, within the embrace of the universal consciousness, and to always find their way home in time for supper.

Surely, the best is yet to come for our magnificent human family, and our thousands of years of evolution and toil will not be wasted. Then, our collective achievement of an alternative future beyond Earth will exceed our every fantasy—a scientific and historical reality so far beyond our existing comprehension, as to be impossible to presently imagine or anticipate.

If we avoid near-term extinction this century, and we continue our natural evolution in harmony with the earth, with the metamorphosis of the minds of humanity bonding with each other and the universal quantum consciousness, we will liberate and empower our children (and their AI robots) to spin from the safe harbor and garden of Earth, while she remains habitable and hospitable to human existence.

Should we fail, however, the inevitable price of continued, willful ignorance will be for our children to experience horribly cruel and painful deaths due to the human-caused reversal of the normal carbon dioxide (CO_2), heat-ice-age cycle.[17]

---

17   Experiencing the glorious, rapid warming and greening of the earth following the most recent 100,000 yearlong ice age, human civilization has flourished all over the world during the last 10,000 pleasant years.

In the normal cycle, Earth should now be very, very slowly cooling toward a distant ice age, providing a long, extended period of a moderate climate; however, the carbon dioxide pumped into the atmosphere by our industry is now reversing the long-term cycle, which, if uncorrected, will result in a runaway burning of Earth not suffered by her for more than 600 million years.

Without its atmosphere and the greenhouse effect provided by CO_2, the earth would be a very cold (and hot) lifeless body like the moon (-208°F in the shade and +250°F in the sunlight); however, Earth's tiny 0.04 percent concentration of CO_2 in the eggshell thin layer of her nitrogen-oxygen atmosphere (if Earth was the size of an egg) provides a very delicate balance between such extreme cold and heat, with little or no margin of safety.

The Global Monitoring Laboratory at Mauna Loa Observatory in Hawaii has seen steady, disastrous, annual increases in the CO_2 levels every year since recordings began in 1958, which are now accelerating every year, as our weather is becoming increasingly unpredictable and dangerous.

Given the current environmental, economic, and political realities, and our instinctive intolerant resistance to diversity and change, it is far more likely than not that the extinction of humanity will become irreversible and certain by the end of this century due to the ignorant denial of human-caused climate change resulting from corporate propaganda and the political suppression of solutions.

Or the end may happen more quickly as a sudden, devastating, worldwide economic collapse tomorrow or next week, or some other completely unforeseen global catastrophe, when our smart phones, computerized bank accounts, property deeds, contracts, paper money, and cryptocurrency become worthless. Complete, worldwide, lawless chaos!

It is critical that we study the lessons of scientific history, as we already know the correct answers and need only to avoid repeating the errors of the past.

We must make better use of our current level of knowledge to improvise and to create new tools to learn more. As we collectively become more observant and aware of self and surroundings, we must effectively engage our collective wisdom to learn new remedies for the instinctual, animalistic, inherited diseases of intolerance, deception, threats, violence, and war that infect the brainstems of every human. Each of us must come to know, understand, and heal our instinctive, congenital intolerance through the acquired immunities of empathy, reason, and tolerance.

Undoubtedly, we have been patiently and lovingly watched, without contact, throughout human evolution, as outside interference would collapse the waveform of our existence by reducing our options, eliminating choices, and curtailing our intellectual creativity.

Left to fend for ourselves, we humans have created our own scientifically historical, unique marvels to behold all around our earth and water garden, and we have generated a healthy reservoir of youthful strength, vigor, and intelligence in our children that is often unappreciated by older generations.

That youthful zeal and playful energy can surprise and amaze parents, as our children somehow succeed in successfully raising responsible grandchildren, who must now assume the burden of ending war and cleaning up the damage to the Garden of Earth resulting from our ignorance and corporate greed. The accumulated waste from industrialization and militarization may be the price paid for our evolution to this point; however, the future of humanity rests within the joyful and creative minds of our young people, and the difficult choices they are now forced to make.

We are rapidly approaching an environmental tipping point where we either overcome the instinctive brainstem intolerances that plague all of us, and we collectively work out our survival together, *or* else everything our human society has labored to build during the pleasant weather of the last 10,000 years will quickly disintegrate into oblivion.

No one, no matter how wealthy or powerful, can survive alone. We will either make it together, or none of us will.

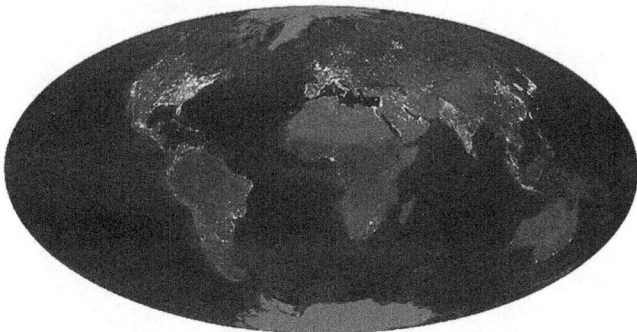

The City Lights of Earth as Observed
from Space (NASA)

Choices

The critical question is whether we humans, who jointly inhabit this earthly garden, can unite to exercise our inherent, universal *Rights of Liberty* to improve our self-governments to better serve our survival and the reality of science. Or do we collapse back into the totalitarianism, nationalism, militarism, and criminal abuse of science and technology that horribly and casually slaughtered millions of children in the last century, and which continue today in Gaza and Ukraine.

The atrocity of conscienceless infanticide in modern technological warfare continues as a calculated byproduct of mindless, industrialized, AI- and computer-directed missiles and drone warfare directed against targets occupied by mothers and their children. This practice is an abomination that threatens the very foundation and definition of our human society.

An apocalyptic war fought with nuclear weapons, missiles, and drones, with computerized AI targeting, will collapse our evolution. Reversing direction, humanity will quickly revert back to tribalism, superstition, and isolation, with the loss of our international law-based society. As our stores and natural resources are exhausted and survival knowledge is lost forever, the inevitable death, destruction, and extinction of humanity will surely occur, with very little evidence of our struggle or achievements to be found in the fossil record millions of years from now.

The primary thing left for observation in such a distant future will be the destructive environmental consequences of our collective existence, which may have already become irreversible. Our era has been labeled the Anthropocene, or Age of Man, as it refers to the geological epoch during which we humans have significantly impacted Earth's climate and ecosystems.

All of us here on Earth—speaking all languages, members of every race, religion, culture, and occupied habitat—are immediately faced with some very real and difficult choices. Each of us, individually, have the universal Rights of Liberty and the innate ability to make our own decisions in life, which creates a power and duty to act to protect the inherent rights demonstrated by our vote—the physical expression of our conditional consent to be governed.

The essential need for collective action is becoming critical at this moment in human history when the most politically vulnerable among us are specifically targeted, effectively conned, and callously victimized to voluntarily vote to surrender our most basic and treasured universal Rights of Liberty.

All around the modern world of politics, from local to international, voters everywhere are being deceptively misled in elections manipulated by paid commercial advertising, where good, honest, and hard-working people are victimized by being intentionally and fraudulently made ever more fearful and angrier by the lies of corrupt politicians to get elected. Most dangerous is the anti-science nature of these misleading campaigns.

Corporate news media outlets of opposing viewpoints peddle inflammatory commentary repeating these political lies to their targeted audiences as truth-challenged entertainment and slanted propaganda to make money through false and misleading advertising.

The purpose of government created by the People's consent has been diverted from the needs of the People, to serve extraconstitutional corporations, their CEOs, and the wealthy, powerful individuals whose bribery finances fraudulent election campaigns. Once elected, representatives do not serve us who cast the ballots; rather they pander to the corporations and the wealthy who profit from wars—instead of those of us who shed our blood on the battlefield and whose tax money to pay for military violence is withheld from our weekly paychecks.

The bipartisan failure of reliable, essential, political information is pandemic, and the misleading accusations and lies by the demagogues of every political persuasion intentionally frighten people into casting instinctive votes of fear. Terror, and the intolerance and hatred it encourages, empowers these authoritarian, lying, and bullying con artists, who promise to save the People from themselves by more forcefully ruling their government.

These reincarnated puffed up strong men, reminiscent of those from the last century, promise to serve as monarchical dictators, operating outside the laws and making decisions for everyone, to combat contrived threats. These despots waste the labor, taxes, lives, and liberties of the People to fight the economic and military wars they start to remain in power.[18]

One personal choice is to continue to mindlessly react to the instinctive warning fears that protectively reside within our reptilian brainstem, causing us to lash out angrily at others who appear different than us, or who threaten our comfort and security.

Or do we evolve our minds to calmly and thoughtfully evaluate and overcome contrived apprehensions, and to find strength in responding positively to the imagination, creativity, and courage of all other minds, as we expand our reality and influence? We can learn to define and magnify our personal, unique, and individual quantum soul as a scientific fact, rather than a religious or philosophical belief.

Do we continue to nourish and evolve the learned tolerance and empathy of our forebrain and to enjoy the benefits of cooperation? Can we purposefully immunize ourselves against the instinctive alarm of intolerance hardwired into our

---

18   Simultaneously published in response to the worldwide crisis in self-government is the companion book *With Liberty and Justice for All* (2025).

brainstem that kept us alive throughout millions of years of competition for food and water?

We can choose to embrace the strengths of individuality and diversity, empowering our fleet minds to soar beyond the reach of our bodily senses and the strictures of quantum gravity. As we identify and pinpoint our special place and document our unique scientifically historical experience within the infinite universe, we can comprehend the unity of everything and the absence of time, absorb the reality of nothingness, and contemplate the implications of an infinite universal consciousness and life after death.

The Age of Sophia Nous

The metamorphosis of the minds of humanity is an expression of the will of the People, as quantum minds generate the powerful *Wisdom of the Crowd* that results from freely collaborating and self-governing minds. Individually, we willfully evolve our minds to peacefully live tolerant and creative lives of joyful reality, rather than to allow our thinking to be infected by ignorant fantasies of wars among the stars and the visceral thrill of killing perceived enemies.

Significant travel beyond the environs of Earth requires the exponential power produced by a shared appreciation and rigorous application of scientific and historical reality. The strength of unity and the absence of conflict will embolden and empower our children to confidently spin from our earthly garden, without fear, mapping the observable regions of the limitless universe in a search for, and visits to, these rare and precious life- and mind-generating, pale blue and green dots of light faintly appearing in the darkness.

These rare planets are gardens of biblical proportions, as they may be the only places within grandmother galaxies where intelligent organic life with a potential for observing quantum minds, can naturally evolve. With social cooperation, we gained the leisure to study the mechanics of objects in the sky, and the curiosity and courage to question the prevalent myths, superstitions, and religious beliefs about how the universe (and humanity) was created and is related—all in a mindful search for the universal truths of scientific and historical reality.

As we emerged from the last ice age into the warming and greening Garden of Earth, our intelligent human forebears were already tracking the movement of heavenly bodies. We quickly evolved as we arose from the megalithic ruins of an advanced peaceful civilization that appears to have existed and thrived in the lower and equatorial latitudes during the Ice Age, and which disappeared as the Age ended. Around the world, we began to reimagine and remember our origins and destiny among the stars that shine in the night sky, which we learned to rely upon for navigating the deserts and seas in our voyages and migrations, as we re-explored our Earth.

Once our cooperation and collective intelligence provided a sufficient livelihood, we could remain in one place long enough to track the brilliant sun that moves across the morning and evening horizons during the year. Identifying the seasons and cycles of life, we built monuments of enduring stone incorporating our knowledge, which we left for our descendants to ponder.

The Göbekli Tepe archeological site in Turkey is the oldest known permanent human settlement in the world. It was built and occupied between 9500 BCE and 8000 BCE, and it includes carvings of "handbags" consisting of a rectangle with a curved handle on top. Thought to perhaps represent the observable earth and curved sky, under superior guidance, the same images have been found at various sites around the world, including an Assyrian relief carving dated 883-859 BCE and an Olmec sculpture dated 1200-400 BCE.

In Greek, *sophia* is a feminine noun meaning wisdom, and *nous* is a masculine noun meaning mind or intellect. Instead of the desolate Age of Man, the combination of *sophia* and *nous* may come to symbolize the stage of human metamorphosis into wise beings of mind. An alternative "Age of Sophia Nous" may be characterized by the individual and collective awareness and observations of mind required to exercise consistent critical thinking, to reason, to make wise decisions, and to act with courage and justice in creating safe and joyful societies.

Concurrent scientific revolutions in cosmology *and* consciousness are revealing the quantum nature of both universe and mind. Exposed is the composite scientific reality of an infinite, living electromagnetic-plasma universe, a cosmos primarily consisting of endless grandmother galaxies, their daughter stars, and the ultimate, natural flowering of organic life with quantum minds that entangle with the universal consciousness following birth and awareness.

It is a privilege and challenge to live at this epochal moment in the evolution of our quantum minds, as we individually become aware that the conscious mind produced by our brain encompasses everything we can perceive as a scientific and historical reality.

Our collective human consciousness generates a marvelous array of imaginable alternatives, all but one of which would cease to exist, should we be interfered with, rather than being lovingly watched by our cosmic relatives. We are never alone, but we alone must create our personal destiny, for such is the way of evolving life in the infinite universe, and the flowering of creative quantum minds.

Every person, every living soul, is accountable for their words and deeds to themselves, to others, and to their memory in the historical reality.

The creations of our imaginations are individual and forever unique in the scientific history of infinity, and our individual efforts are magnified exponentially when combined with others. Acting together, we can peacefully transform the survival drive of our instinctual intolerance into empathetic tolerance and acceptance of the inherent, universal Rights of Liberty of every person to be free of restraint upon their physical existence, self-governance, migration, and creations.

Now is when we will either learn everything there is to know, or we will soon forget everything we have learned.

Appendix A
About Quantumness

These writings propose that the infinite electromagnetic-plasma universe of light is governed by *quantum relativity*, and that the *quantum minds* of the organic intelligent life it gives birth to are also *quantumly entangled* with an infinite cosmic consciousness. An understanding of its modern meaning might benefit from a brief note about the strange *"quantum"* concept in physics and consciousness.

The gravitational theory according to Newton and mathematically modified by Einstein more than a hundred years ago attempted to describe the mechanics of the universe at a time when their observations were limited to the stars of our local Milky Way galaxy. Improved observations since then have repeatedly falsified the theory requiring the creative invention of inflation, black holes, dark matter, and dark energy to salvage it.

Moreover, when going from the very large to the very small, gravitational theory completely fails to describe observations and the relationships of minute particles of mass and massless electromagnetic phenomena.

Under the quantum theory, light is an electromagnetic wave resulting from oscillating electric and magnetic fields. In a duality, all matter is made up of discrete wavelike particles, such as electrons and protons, and its waves impact and perform as individual photons or packets of energy. All these particles of light and matter possess their own unique spin, or angular momentum, identified by their discrete *quanta* of energy.

Quantum states, in which two separated particles become bound, is a state of knowledge or imagination, in that the particles instantaneously "know" the spin of each other, irrespective of the distance between them. Affecting one instantly affects the other. This timeless entanglement creates a state of contemporaneous awareness, rather than an objective physical relationship.

Like mind and time, quantumness is incorporeal, without form or substance. Imperceptible quantum entanglement manifests itself in what we can physically observe and measure, such as gravity and electromagnetism.

Quantum mechanics provides the ability to calculate the properties and behaviors of weightless microscopic physical systems. It cannot predict with certainty what will happen, but it can establish the probability of occurrences with a high degree of confidence.

Once it is understood and accepted that the physical universe is not expanding, that it is infinite, without beginning or end, we can then reinterpret (and finally understand) what we observe with our instruments. Reality is an infinite process of electromagnetism within an eternal plasma that generates endless, living grandmother galaxies, all quantumly related, timelessly, forever and always.

Galaxies ultimately give rise to warm mother fusion stars with rare, fecund earth and water planets with a large moon that creates organic life, which evolves intelligence that acquires language, and becomes aware of itself and its surroundings. Finally, quantum minds arise that can overcome instinctive brainstem intolerance and violence, allowing the cooperative development of the essential wisdom, knowledge, and technology needed to spin, freely, throughout the universe.

Appendix B

The Ratio of Everything to Nothingness

In a thought experiment, the ratio between the negative nothingness and the positive, infinite universe and plasma can be imagined by the difference in volume of the single hydrogen atom that exists within the nothingness of every square meter of intergalactic space, and the volume of the whole cube. The spherical volume surrounding a single, spinning neutral hydrogen atom, with its positive proton orbited by its captured negative electron is approximately 6.2×10^{-31} cubic meters.

If stacked together, without overcrowding, there should be individual spherical spaces within each square meter of intergalactic nothingness for approximately 62,000,000,000,000,000,000 ,000,000,000,000,000 hydrogen atoms, instead of the one atom found there. Moreover, there is the remaining nothingness within each tiny spherical volume between the infinitesimal wave-like particles of protons and electrons, and in the remaining corner spaces between the stacked spheres.

All things considered, the ratio between everything positive, the eternal plasma with every particle of the infinite physical universe, and the encompassing negative nothingness would appear to be near zero, less than 00.01% positive and more than 99.99% negative nothingness.

If we imagine the structure of the observable, positive, infinite, physical universe defined by a mathematics based on the geometric progression of the powers of two, we might also visualize that structure contained by an organization of the negative nothingness resulting from the geometric progression of the powers of three. This is demonstrated by the elemental plasma crystal consisting of 3^2 free electrons, 3^2 naked protons, and 3^3 hydrogen atoms, all vibrating in electrostatic harmony.

Appendix C

A True Story About
the Family of Jesus

About 2,000 years ago, a truly amazing family lived in Northern Israel. The Land was ruled as a police state by the incestuous Herodian monarchy at the sufferance of the Roman Empire and its occupying army. The roads in the Galilee were lined with the crucified corpses of freedom fighters, left hanging by the Romans as rotting examples of the futility of resistance to their Empire.[19]

Living with the daily terror, the family, as did most ordinary people at the time, followed the Way of the Osim (Essene), as simple "Doers of the Law." The "poor" followers of this uncomplicated expression of liberty and religious freedom had lived peaceful lives of righteousness and justice for hundreds of years throughout the Land, since before its conquest by Alexander the Great and administration by the Greeks, the subsequent Hasmonean rule, and defeat by the Roman army.

---

19 *The Way of Righteousness* concluded with a brief story about the family of Jesus, which is included here. With more than half of the world population consisting of Christians and Muslims influenced by the essential message of Jesus, the historical reality of his life is relevant to a comprehension of the scientific reality of an infinite universe.

The Osim built a refuge in the Judean wilderness at Qumran on the shore of the Dead Sea, where they sought solitude to study and reflect upon their books that promised not one, but three messiahs, a suffering Son of Man, a Davidic Leader, and a Priestly Messiah, all of whom would come to rescue the people from the evil Herodian monarchy and Roman empire that ruled their lives.

Cleopas (Biblical–Joseph), the father of the Galilean family, was a Rechabite,[20] who traveled to find work as a carpenter, while the mother, Mary (Aramaic-Maryam, Hebrew-Miryam), delivered and raised five sons and a daughter. Her first born were twins, Yeshua (Greek-Jesus) and his brother, Judah (Greek-Judas, or Thomas-Aramaic for twin). The twins became two of the three expected messiahs of the Osim, and the next of Mary's sons, Jacob (Greek-James), became the third messiah.

At their births, Mary and Cleopas dedicated the firstborn of her twins, Jesus, and the following two of their sons, James and Simeon, as Nazarites.[21]

---

20 The Rechabites were a clan of the Kenites, who arrived from Egypt with the Mosaic priesthood and settled among the Israelites. The Kenites avoided alcohol, and engaged in highly skilled metal and wood working, traveling between towns and cities.

21 Nazarites lived the lives they were consecrated to, for so long as they lived. Their time was dedicated to study and contemplation, as they abstained from alcohol, cutting their hair unnecessarily, oiling their bodies, getting married, or fathering children. (Numbers 6:8).

The boys were raised to live, every day of their lives, from birth to death, as being "holy unto the Lord." At age 10, each of the Nazarite boys left home to prepare for the alternative priesthood of the Osim, the Sons of Zadok.

One by one, the brothers came to live in isolation at the Qumran refuge, to seek wisdom in its libraries and self-awareness during solitary reflection, to identify their mission in life, and to experience the vision of the path they would follow.

After twenty years of study, first Jesus, then James, and then Simeon achieved intellectual maturity at age thirty, and were ordained as priests of the Sons of Zadok. The brothers were prepared to teach the Way of Righteousness, and to spiritually lead the people and their zealous Sons of Light in both war and peace.

When the sun rose over the Dead Sea on their thirtieth birthday, each brother left the serenity of the Osim refuge and walked the harsh 29-mile winding path up through the hills of the Judean wilderness, past the village of Bethany, to the gates of Jerusalem. Each brother must have fully contemplated the immense cruelty that awaited him, and the risks and consequences of the journey that lay ahead.

All five of Mary's sons died violently, sacrificed by their loving parents on the altar of eternal liberty, in fulfillment of their covenant of righteousness and justice.

The Osim and Their Way of Righteousness

The Osim simply abided by the ancient covenant of Abraham to peacefully live out each moment of life in righteousness and justice. These "Doers of the Law" had no respect for the oral law of the lawyerly Pharisees and their "fences around the law," nor could they tolerate the corrupt polluting practices of the Sadducean priesthood at the Temple, both of whom collaborated with and spied for the Herodians and their Roman masters in suppressing the liberty of the suffering people. The "Poor" had been separated from the priests and lawyers for centuries, as the elites had collaborated with the Greek invaders before the Romans.

The Osim, and their Way of Righteousness, rebelled against the brutal repression. The guerilla war for liberty was fought by their Zealot warriors—the Sons of Light; it was spiritually led by their priests—the Sons of Zadok; and the wounded, widows, and orphans were comforted by their lay ministers—the Order of Melchizedek.

As Jerusalem fell to the Roman army—and the children of the Way were being rounded up throughout the Land to be sold into slavery in the Empire, and the most defiant were thrown to the wild animals in the amphitheater at Caesarea by the Herodians and Romans in celebration of their victory—the Osim concealed their books in sealed jars within remote caves among the cliffs around Qumran by the Dead Sea.

Their archives included Cave Four, which was created by connecting the interiors of several adjacent caves located high up the face of a cliff across from the Refuge. The Osim enlarged and squared out a large cave behind the small entrances, which were accessible only by ropes and ladders. The large library provided shelf space to organize a thousand scrolls, and a peaceful, cool place for

boys and men to read, reflect, talk, imagine their futures, and to think about what each could and must do to contend with the evil power of empire.

The Zealot warriors of the Way continued to resist the Roman army, the mightiest in the world, even after Jerusalem and Qumran fell. The Sons of Light retreated to Herod's massive desert fortress on top of the mountain at Masada, further south in the desert wilderness. Herod had been besieged at Masada by the Zealots early in his rule—until he was rescued by the Roman army. He continued to expand his palaces at the fortress throughout his reign, but Masada was captured in a surprise attack by the Zealots during the war.

After fighting off the Roman army for four years, and as the gate was collapsing, the Sons of Light buried their last remaining books of the Way of Righteousness, drew lots, and died by their own swords, rather than by the weapons wielded by the brutal hands of empire.

The Children of Mary

Jesus became the Osim Suffering Son of Man Messiah predicted by Isaiah and expected by the Way of Righteousness. He cleansed the Temple of Sadducean pollution, caused a riot, and he was arrested and summarily executed by the Romans.

Jesus sacrificed his life for the salvation of his people from the cruelty, corruption, and power of the Herodians and Romans. In doing so, he liberated his soul, so his words of righteousness would live on in the minds of all those who seek the truth of reality.

Judas Thomas, the twin brother of Jesus, was not a Nazirite, and he fathered a family. Armed with a sword, Thomas became the Osim's Davidic Messiah. He took the message of the Way of Righteousness to the East, where the Way guided the Ebionite Christian kingdoms established there and which existed for hundreds of years. Thomas extended the Way into Syria, Saudi Arabia, Iraq, Iran, and finally India, where he was struck down by a spear and died with his sword in his hand.

Thomas's Letter of Jude concludes the New Testament, just before Revelations, and his original Gospel of Thomas (revealed in the Gnostic Gospels) was used in the fabrication of the Pauline Synoptic Gospels.

Subsequent Roman emperors felt threatened by the lineal descendants of Judas Thomas, whom they feared might create a unified Davidic kingdom in the Middle East. For more than five centuries, Judas's teachings were recited and taught in these lands, until finding eloquent expression in Saudi

Arabia by Muhammad, the last prophet of Allah, in his Islamic message of righteousness.

Mary's third son, and second Nazirite, was Jacob, or James, who became the Priestly Messiah predicted by the Osim. Following the crucifixion of Jesus by the Romans, James the Just led the Way and the Zadok priesthood for 26 years in Jerusalem and Qumran. His essential teaching about works versus faith is contained in the New Testament Letter of James and the Dead Sea Scrolls.

As a Righteous Teacher of the Way, and as a priest of the alternative Zadok priesthood, James the Just was elected by the People and the Sons of Zadok, to represent them as their High Priest in the Temple cleansed by Jesus. For this heresy, a Pharisaic mob led by the enigmatic Saul (Paul), threw James down the Temple steps, breaking both of his legs. After recovering at Qumran and returning to Jerusalem, James was stoned to death near the Mount Zion cornerstone of the Temple, as the result of a conspiracy by the Herodians, Sadducees, and Pharisees, to whom James and the Way posed a threat to their power over the People.

Following the execution of James, his younger brother—the third of Mary's Nazirite sons—Simeon bar Cleopas, a Rechabite priest of the Sons of

Zadok, was elected leader of the Way. After the Romans conquered the defending Zealots, leveled Jerusalem, and were advancing to destroy the refuge at Qumran, the Osim concealed their books in caves and Simeon led their escape across the Jordan River. As other zealous Sons of Light remained to fight the Romans, Simeon returned as their spiritual leader. He was ultimately captured and crucified by the Romans near Jerusalem.

The youngest of Mary's sons, Joseph, who is also known in the New Testament as Joses, or Barnabas (Hebrew—son of comfort), probably wrote the original Book of Matthew in Hebrew, before writing his Letter to the Hebrews in the Greek language. The mission of Joseph, as a lay minister of the Order of Melchizedek, was to teach the works of righteousness to the Gentiles, to whom the Order was open. (Hebrews 7:15-22)

Joseph was assigned by James to supervise the ministry of Paul (who claimed conversion to the Osim and spent three years studying at Qumran). Paul was instructed to teach the Way in his hometown of Tarsus in Asia Minor; however, Paul went off on his own in defiance of James and created a Pharisaic Christianity, which supported the Romans and the God-given rule of kings and emperors.

It is said that Joseph was either beheaded or burned to death by the Romans on the island of Cyprus. That may be true, but others believe Joseph escaped to Ephesus to edit the Gospel of John in his old age. The Gospel is a revision of the Roman Catholic Synoptic Gospels to better reconcile them with Mary Magdalene's Gnostic teaching of the Spirit of Wisdom in the West, with James's words of righteousness at Qumran, and with the message of Judas Thomas in the East.

The only known daughter of Mary was named Salome. She was present at Jesus's crucifixion, helped recover his body, and, with others, she accompanied it back through the wilderness to Qumran. There, she helped wash and wrap Jesus's body for burial, as his grave was being dug in the nearby cemetery and rocks were gathered to cover it. Salome lived on to care for her parents in their old age, and to attend to their deaths.

The bodies of Jesus, James, Simeon and others of his family are probably buried there, somewhere, under piles of stones in the cemeteries of the Osim at Qumran, forgotten in the sun and rain for 2,000 years, until now.

Mary Magdalene, the Companion

Mary Magdalene, her sister Martha, and their brother Lazarus were followers of the Way. They lived in Bethany, near Jerusalem on the southeastern slope of the Mount of Olives, along the road coming up out of the wilderness from Jericho and Qumran. As Jesus rested overnight, on his way to cleanse the Temple and certain death, Mary Magdalene anointed him to be the Way's Suffering Son of Man Messiah. Mary was Jesus's favored companion in life, and she was the heir to his most spiritual and esoteric teachings.

Mary Magdalene took the Way's Spirit of Wisdom to the West. She and her Gnostic followers taught the enlightened message of Jesus and his Way of Righteousness for hundreds of years throughout Egypt, Syria, Asia Minor, Greece, and into France. Gnosticism remained the prevalent study of the original teachings of Jesus (without his being God), until Emperor Constantine seized Roman power in the fourth century, and he designated Paul's Roman Catholic Church as the Empire's only lawful expression of Christianity. Jesus was made out to be the son of God,[22] united with God and the Holy Ghost (adapted from the Way's Spirit of Wisdom) as a Trinity that recognized and encouraged the earthly reign of kings and empires. Original sin was invented, and Catholic priests were empowered to forgive.

Mary is honored throughout the Gnostic Gospels as the one Jesus loved the most. How long she lived, how far she traveled, and how and where she died are unknown. Some believe her remains may be buried somewhere in southern France.

Mary's Spirit of Wisdom was freed when the jars containing the Gnostic Gospels and the Dead Sea Scrolls were broken open at the end of World War II, and the written records of the distant past

---

22 Thousands of Mary's Gnostics were burned at the stake during the Catholic Inquisitions, rather than to acknowledge Jesus as God, yet their spiritual belief in the essential message of the historical Jesus lives on to this day in the minds of many, if not most Christians who accept the essence of his message of love, and his suffering for the salvation of the people from the cruelty of empire and war.

were revealed in our lifetime to be read and considered, once again, after 2,000 years of concealment.

The Revelation of the Books

As the libraries of the Gnostics were being seized and burned by Roman Catholic Church authorities throughout the empire in the fourth century, 52 books were sealed in a large jar buried at the base of a cliff in Egypt, near the Nile River. The books remained there, undisturbed, for 1,700 years, until 1945.[23]

Within a year of the Gnostic Gospels being unburied in Egypt, the Dead Sea Scrolls were discovered in 1946, also preserved in jars, in a cave at Qumran, a few hundred miles to the east. Less than 20 years later, other books of the Way were recovered from under the floor of the meeting chamber at Masada. The books had been buried there by the last of the Zealot Sons of Light in 73 CE.

The courageous cultural and spiritual warriors of the Way defended their families, their books, and their universal rights of liberty to live their

---

23 Mary's Spirit of Wisdom survived being mischaracterized as the Holy Ghost, the Catholic Inquisition and Crusades, and the Catholic-Protestant Wars, until now. As religious and cultural wars rage on in the Middle East, and babies are being slaughtered in Gaza, the Spirit of Wisdom can still be heard upon the winds of time, whispering words of comfort, and weeping for her children.

lives in peace, righteousness, and justice as promised by Abraham when he sought to sojourn in the Land of Israel. Their resistance was crushed by the cruel and ignorant power of empire—which has continued to compete for world domination for 2,000 years. Until now, when the courage of those spiritual warriors who fought and died for the liberty of their people, was finally revealed by the books they read and wrote.

With their minds at ease, and with the Roman army battering against the great gate Herod had built at Masada, the zealous Sons of Light did not fear final judgment. They were at one with themselves, and with others of the Way, with whom they had lived, fought, and died for the right to peacefully live simple lives of liberty and justice.

Nor did the warriors of the Way fear the pain of death or worry where their corpses might be thrown to decay. For them, their souls were set free over these thousands of years, to survive death in the minds of all those who remember them and their courage.

The bravery of the Sons of Light can inspire us to once again confront the inherent evil of empires, the ignorant wars they fight, and the children they wantonly slaughter, without conscience, remorse, or expectation of accountability.

The miraculous discovery of these hidden books revealed the scientific history of who these extraordinary people really were, the meaning of their lives of liberty, and the truth about what they lived and died for. Today, we can draw upon the power of their wisdom and courage to help us confront the environmental, economic, political, military, and intolerance crises of our time and to avoid our imminent extinction within the lifetime of babies being born today.

Sources

Aguilar, David A., "Gamma-Ray Jets from the Milky Way," (Science Update, Center for Astrophysics, Harvard & Smithsonian, 05.31.12).

Alfvén, Hannes, *Cosmic Plasma*, (D. Reidel Publishing, 1981).

Arp, Halton, *Catalogue of Discordant Redshift Associations*, (Apeiron, 2003).

Arp, Halton, *Seeing Red: Redshifts, Cosmology and Academic Science*, (Apeiron, 1998).

Ashmore, Lyndon E., "Calculating the redshifts of distant galaxies from first principles by the new tired light theory (NTL)," (Journal of Physics: Conference Series, (2019 J.Phys. Conf Ser 1251 012007).

Ashmore, Lyndon E., *Tired Light: an explanation of redshifts in a static universe*, (Independent, 2016).

Ashmore, Lyndon E., *Big Bang Blasted! The story of the expanding universe and how it was shown to be wrong*, (BookSurge Publishing, 2006).

Aslan, Reza, Zealot: The Life and Times of Jesus of Nazareth, (Random House, 2013).

Borchardt, Glenn, *Infinite Universe Theory*, (Progressive Science Institute, 2017).

Brubaker, Ben, "Computer Scientists Prove That Heat Destroys Quantum Entanglement," (*Quanta Magazine*, August 28, 2024).

Eisenman, Robert, *James the Brother of Jesus: The Key to Unlocking the Secrets of Early Christianity and the Dead Sea Scrolls*, (Viking, 1996).

Eisenman, Robert & Wise, Michael, *The Dead Sea Scrolls Uncovered*, (Element, 1992).

Findlay, Tom, *A Beginner's View of Our Electric Universe*, (Grosvenor House of Publishing, 2013).

Finkelstein, Israel & Silberman, Neil Asher, *The Bible Unearthed: Archaeology's New Vision of Ancient Israel and the Origin of Its Sacred Texts*, The Free Press, 2001).

Fisher, Deanne, "New measurements reveal the enormous halos that shroud all galaxies in the universe," (The Conversation, September 6, 2024).

Friedman, Richard Elliott, *Who Wrote the Bible?*, (Harper & Row,1987).

Hobson, Art, *Tales of the Quantum: Understanding Physics' Most Fundamental Theory*, (Oxford University Press, 2017).

Sources

Hoyle, Fred, Burbidge, et al., *A Different Approach to Cosmology: From a Static Universe Through the Big Bang Towards Reality*, (Cambridge University Press, 2000).

Kakalios, James, *The Amazing Story of Quantum Mechanics*, (Gotham Books, 2010).

Lerner, Eric J., *The Big Bang Never Happened: A Startling Refutation of the Dominant Theory of the Origin of the Universe*, (Vintage Books, 1992).

Lindley, David, *The Dream Universe: How Fundamental Physics Lost its Way*, (Doubleday, 2020).

Mamas, Dean L., "An explanation for the cosmological redshift," (*Physics Essays*, 2010).

Marmet, Louis, editor, *A Cosmology Group*, cosmology.info).

McFadden, Johnjoe, *Quantum Evolution: Life in the Multiverse*, (Flamingo, 2000).

Michael, George, *Willis Carto and the American Far Right*, (University Press of Florida, 2008).

Pagels, Elaine, *The Gnostic Gospels*, (Vintage, 1979).

Peebles, P.J.E., *Principles of Physical Cosmology*, (Princeton University Press, 1993).

Penrose, Sir Roger; Hameroff, Stuart, et al., *Consciousness and the Universe: Quantum Physics, Evolution, Brain & Mind*, (Science Publishers, 2017).

Peratt, Anthony L., *Physics of the Plasma Universe*, (Springer, Second Edition, 2015).

Rae, Alastair I. M., *Quantum Physics: A Beginner's Guide*, (One World, 2005).

Ratcliffe, Hilton, *The Static Universe: Exploding the Myth of Cosmic Expansion*, (C. Roy Keys, 2010).

Robinson, James M. Ed., *The Nag Hammadi Library*, (Harper & Row, 1977).

Rosenblum, Bruce & Kuttner, Fred, *Quantum Enigma: Physics Encounters Consciousness*, (Oxford, 2011).

Sanejouand, Yves-Henri, "A framework for the next generation of stationary cosmological models," (Facult´e des Sciences et des Techniques, Nantes, France, December 21, 2021).

Schneider, Peter, *Extragalactic Astronomy and Cosmology: An Introduction*, (Springer, Second Edition, 2015).

Scott, Donald E., *The Electric Sky: A Challenge to the Myths of Modern Astronomy*, (Mikamar Publishing, Second Edition, 2012).

Scott, Donald E., *The Interconnected Cosmos*, (Stickmanonstone Publishing, 2021).

Shao, Ming-Hui, "The energy loss of photons and cosmological redshift," (*Physical Essays*, 2013).

Silberman, Neil Asher, *The Hidden Scrolls: Christianity, Judaism, and the War for the Dead Sea Scrolls*, Grosset/Putnam, 1994).

Steinhardt, Paul J. and Turok, Neil, *Endless Universe: Beyond the Big Bang*, (Doubleday, 2007).

Surowiecki, James, *The Wisdom of Crowds: Why the Many Are Smarter Than the Few and How Collective Wisdom Shapes Business*, (Doubleday, 2004).

Susskind, Leonard & Friedman, Art, *Quantum Mechanics: The Theoretical Minimum*, (Basic Books, 2014).

Thornhill, Wallace and Talbott, David, *The Electric Universe: A New View of Earth, the Sun, and the Heavens*, (Mikamar Publishing, 2008).

Made in the USA
Monee, IL
30 August 2025

24620999R00075